Raoul Huys and Viktor K. Jirsa (Eds.)

Nonlinear Dynamics in Human Behavior

Studies in Computational Intelligence, Volume 328

Editor-in-Chief
Prof. Janusz Kacprzyk
Systems Research Institute
Polish Academy of Sciences
ul. Newelska 6
01-447 Warsaw
Poland
E-mail: kacprzyk@ibspan.waw.pl

Further volumes of this series can be found on our homepage: springer.com

Vol. 306. Tru Hoang Cao
Conceptual Graphs and Fuzzy Logic, 2010
ISBN 978-3-642-14086-0

Vol. 307. Anupam Shukla, Ritu Tiwari, and Rahul Kala
Towards Hybrid and Adaptive Computing, 2010
ISBN 978-3-642-14343-4

Vol. 308. Roger Nkambou, Jacqueline Bourdeau, and Riichiro Mizoguchi (Eds.)
Advances in Intelligent Tutoring Systems, 2010
ISBN 978-3-642-14362-5

Vol. 309. Isabelle Bichindaritz, Lakhmi C. Jain, Sachin Vaidya, and Ashlesha Jain (Eds.)
Computational Intelligence in Healthcare 4, 2010
ISBN 978-3-642-14463-9

Vol. 310. Dipti Srinivasan and Lakhmi C. Jain (Eds.)
Innovations in Multi-Agent Systems and Applications – 1, 2010
ISBN 978-3-642-14434-9

Vol. 311. Juan D. Velásquez and Lakhmi C. Jain (Eds.)
Advanced Techniques in Web Intelligence, 2010
ISBN 978-3-642-14460-8

Vol. 312. Patricia Melin, Janusz Kacprzyk, and Witold Pedrycz (Eds.)
Soft Computing for Recognition based on Biometrics, 2010
ISBN 978-3-642-15110-1

Vol. 313. Imre J. Rudas, János Fodor, and Janusz Kacprzyk (Eds.)
Computational Intelligence in Engineering, 2010
ISBN 978-3-642-15219-1

Vol. 314. Lorenzo Magnani, Walter Carnielli, and Claudio Pizzi (Eds.)
Model-Based Reasoning in Science and Technology, 2010
ISBN 978-3-642-15222-1

Vol. 315. Mohammad Essaaidi, Michele Malgeri, and Costin Badica (Eds.)
Intelligent Distributed Computing IV, 2010
ISBN 978-3-642-15210-8

Vol. 316. Philipp Wolfrum
Information Routing, Correspondence Finding, and Object Recognition in the Brain, 2010
ISBN 978-3-642-15253-5

Vol. 317. Roger Lee (Ed.)
Computer and Information Science 2010
ISBN 978-3-642-15404-1

Vol. 318. Oscar Castillo, Janusz Kacprzyk, and Witold Pedrycz (Eds.)
Soft Computing for Intelligent Control and Mobile Robotics, 2010
ISBN 978-3-642-15533-8

Vol. 319. Takayuki Ito, Minjie Zhang, Valentin Robu, Shaheen Fatima, Tokuro Matsuo, and Hirofumi Yamaki (Eds.)
Innovations in Agent-Based Complex Automated Negotiations, 2010
ISBN 978-3-642-15611-3

Vol. 320. xxx

Vol. 321. Dimitri Plemenos and Georgios Miaoulis (Eds.)
Intelligent Computer Graphics 2010
ISBN 978-3-642-15689-2

Vol. 322. Bruno Baruque and Emilio Corchado (Eds.)
Fusion Methods for Unsupervised Learning Ensembles, 2010
ISBN 978-3-642-16204-6

Vol. 323. Yingxu Wang, Du Zhang, and Witold Kinsner (Eds.)
Advances in Cognitive Informatics, 2010
ISBN 978-3-642-16082-0

Vol. 324. Alessandro Soro, Vargiu Eloisa, Giuliano Armano, and Gavino Paddeu (Eds.)
Information Retrieval and Mining in Distributed Environments, 2010
ISBN 978-3-642-16088-2

Vol. 325. Quan Bai and Naoki Fukuta (Eds.)
Advances in Practical Multi-Agent Systems, 2010
ISBN 978-3-642-16097-4

Vol. 326. Sheryl Brahnam and Lakhmi C. Jain (Eds.)
Advanced Computational Intelligence Paradigms in Healthcare 5, 2010
ISBN 978-3-642-16094-3

Vol. 327. Slawomir Wiak and Ewa Napieralska-Juszczak (Eds.)
Computational Methods for the Innovative Design of Electrical Devices, 2010
ISBN 978-3-642-16224-7

Vol. 328. Raoul Huys and Viktor K. Jirsa (Eds.)
Nonlinear Dynamics in Human Behavior, 2010
ISBN 978-3-642-16261-9

Raoul Huys and Viktor K. Jirsa (Eds.)

Nonlinear Dynamics in Human Behavior

Raoul Huys
Theoretical Neuroscience Group
CNRS & Université de la Méditerranée, UMR 6233 "Movement Science Institute"
Faculté des Sciences du Sport
163 av. De Luminy
13288, Marseille cedex 09
France
E-mail: raoul.huys@univmed.fr

Viktor K. Jirsa
Theoretical Neuroscience Group
CNRS & Université de la Méditerranée, UMR 6233 "Movement Science Institute"
Faculté des Sciences du Sport
163 av. De Luminy
13288, Marseille cedex 09
France
and
Center for Complex Systems & Brain Sciences
Florida Atlantic University
777 Glades Road
Boca Raton FL33431
USA

ISBN 978-3-642-16261-9   e-ISBN 978-3-642-16262-6

DOI 10.1007/978-3-642-16262-6

Studies in Computational Intelligence   ISSN 1860-949X

Library of Congress Control Number: 2010937350

© 2010 Springer-Verlag Berlin Heidelberg

This work is subject to copyright. All rights are reserved, whether the whole or part of the material is concerned, specifically the rights of translation, reprinting, reuse of illustrations, recitation, broadcasting, reproduction on microfilm or in any other way, and storage in data banks. Duplication of this publication or parts thereof is permitted only under the provisions of the German Copyright Law of September 9, 1965, in its current version, and permission for use must always be obtained from Springer. Violations are liable to prosecution under the German Copyright Law.

The use of general descriptive names, registered names, trademarks, etc. in this publication does not imply, even in the absence of a specific statement, that such names are exempt from the relevant protective laws and regulations and therefore free for general use.

Typeset & Cover Design: Scientific Publishing Services Pvt. Ltd., Chennai, India.

Printed on acid-free paper

9 8 7 6 5 4 3 2 1

springer.com

Preface

In July 2007 the international summer school "Nonlinear Dynamics in Movement and Cognitive Sciences" was held in Marseille, France. The aim of the summer school was to offer students and researchers a "crash course" in the application of nonlinear dynamic system theory to cognitive and behavioural neurosciences. The participants typically had little or no knowledge of nonlinear dynamics and came from a wide range of disciplines including neurosciences, psychology, engineering, mathematics, social sciences and music. The objective was to develop sufficient working knowledge in nonlinear dynamic systems to be able to recognize characteristic key phenomena in experimental time series including phase transitions, multistability, critical fluctuations and slowing down, etc. A second emphasis was placed on the systematic development of functional architectures, which capture the phenomenological dynamics of cognitive and behavioural phenomena. Explicit examples were presented and elaborated in detail, as well as "hands on" explored in laboratory sessions in the afternoon. This compendium can be viewed as an extended offshoot from that summer school and breathes the same spirit: it introduces the basic concepts and tools adhering to deterministic dynamical systems as well as its stochastic counterpart, and contains corresponding applications in the context of motor behaviour as well as visual and auditory perception in a variety of typically (but not solely) human endeavours. The chapters of this volume are written by leading experts in their appropriate fields, reflecting ta similar multi-disciplinary range as the one of the students. This book owes its existence to their contributions, for which we wish to express our gratitude. We are further indebted to the Technical Editor Dr. Thomas Ditzinger for his advice, guidance, and patience throughout the editorial process, and the Series Editor Janusz Kacprzyk for inviting and encouraging us to produce this volume.

Contents

Dynamical Systems in One and Two Dimensions:
A Geometrical Approach 1
Armin Fuchs

Benefits and Pitfalls in Analyzing Noise in Dynamical
Systems – On Stochastic Differential Equations and System
Identification 35
Andreas Daffertshofer

The Dynamical Organization of Limb Movements 69
Raoul Huys

Perspectives on the Dynamic Nature of Coupling in Human
Coordination 91
Sarah Calvin, Viktor K. Jirsa

Do We Need Internal Models for Movement Control? 115
Frédéric Danion

Nonlinear Dynamics in Speech Perception 135
Betty Tuller, Noël Nguyen, Leonardo Lancia, Gautam K. Vallabha

A Neural Basis for Perceptual Dynamics 151
Howard S. Hock, Gregor Schöner

Optical Illusions: Examples for Nonlinear Dynamics in
Perception 179
Thomas Ditzinger

A Dynamical Systems Approach to Musical Tonality 193
Edward W. Large

Author Index 213

Dynamical Systems in One and Two Dimensions: A Geometrical Approach

Armin Fuchs

Abstract. This chapter is intended as an introduction or tutorial to nonlinear dynamical systems in one and two dimensions with an emphasis on keeping the mathematics as elementary as possible. By its nature such an approach does not have the mathematical rigor that can be found in most textbooks dealing with this topic. On the other hand it may allow readers with a less extensive background in math to develop an intuitive understanding of the rich variety of phenomena that can be described and modeled by nonlinear dynamical systems. Even though this chapter does not deal explicitly with applications – except for the modeling of human limb movements with nonlinear oscillators in the last section – it nevertheless provides the basic concepts and modeling strategies all applications are build upon. The chapter is divided into two major parts that deal with one- and two-dimensional systems, respectively. Main emphasis is put on the dynamical features that can be obtained from graphs in phase space and plots of the potential landscape, rather than equations and their solutions. After discussing linear systems in both sections, we apply the knowledge gained to their nonlinear counterparts and introduce the concepts of stability and multistability, bifurcation types and hysteresis, hetero- and homoclinic orbits as well as limit cycles, and elaborate on the role of nonlinear terms in oscillators.

1 One-Dimensional Dynamical Systems

The one-dimensional dynamical systems we are dealing with here are systems that can be written in the form

$$\frac{dx(t)}{dt} = \dot{x}(t) = f[x(t), \{\lambda\}] \tag{1}$$

In (1) $x(t)$ is a *function*, which, as indicated by its argument, depends on the *variable* t representing time. The left and middle part of (1) are two ways of expressing

---

Armin Fuchs
Center for Complex Systems & Brain Sciences, Department of Physics,
Florida Atlantic University
e-mail: fuchs@ccs.fau.edu

how the function $x(t)$ changes when its variable t is varied, in mathematical terms called the *derivative* of $x(t)$ with respect to t. The notation in the middle part, with a dot on top of the variable, $\dot{x}(t)$, is used in physics as a short form of a derivative with respect to time. The right-hand side of (1), $f[x(t),\{\lambda\}]$, can be any function of $x(t)$ but we will restrict ourselves to cases where f is a low-order polynomial or trigonometric function of $x(t)$. Finally, $\{\lambda\}$ represents a set of *parameters* that allow for controlling the system's dynamical properties. So far we have explicitly spelled out the function with its argument, from now on we shall drop the latter in order to simplify the notation. However, we always have to keep in mind that $x = x(t)$ is not simply a variable but a function of time.

In common terminology (1) is an ordinary autonomous differential equation of first order. It is a *differential equation* because it represents a relation between a function (here x) and its derivatives (here $\dot{x}$). It is called *ordinary* because it contains derivatives only with respect to one variable (here t) in contrast to *partial* differential equations that have derivatives to more than one variable – spatial coordinates in addition to time, for instance – which are much more difficult to deal with and not of our concern here. Equation (1) is *autonomous* because on its right-hand side the variable t does not appear explicitly. Systems that have an explicit dependence on time are called *non-autonomous* or *driven*. Finally, the equation is of *first order* because it only contains a first derivative with respect to t; we shall discuss second order systems in sect. 2.

It should be pointed out that (1) is by no means the most general one-dimensional dynamical system one can think of. As already mentioned, it does not explicitly depend on time, which can also be interpreted as decoupled from any environment, hence autonomous. Equally important, the change $\dot{x}$ at a given time t only depends on the state of the system at the same time $x(t)$, not at a state in its past $x(t-\tau)$ or its future $x(t+\tau)$. Whereas the latter is quite peculiar because such systems would violate causality, one of the most basic principles in physics, the former simply means that system has a memory of its past. We shall not deal with such systems here; in all our cases the change in a system will only depend on its current state, a property called *markovian*.

A function $x(t)$ which satisfies (1) is called a *solution* of the differential equation. As we shall see below there is never a single solution but always infinitely many and all of them together built up the *general solution*. For most nonlinear differential equations it is not possible to write down the general solution in a closed analytical form, which is the bad news. The good news, however, is that there are easy ways to figure out the dynamical properties and to obtain a good understanding of the possible solutions without doing sophisticated math or solving any equations.

1.1 Linear Systems

The only linear one-dimensional system that is relevant is the equation of continuous growth

$$\dot{x} = \lambda x \tag{2}$$

where the change in the system $\dot{x}$ is proportional to state x. For example, the more members of a given species exist, the more offsprings they produce and the faster the population grows given an environment with unlimited resources. If we want to know the time dependence of this growth explicitly, we have to find the solutions of (2), which can be done mathematically but in this case it is even easier to make an educated guess and then verify its correctness. To solve (2) we have to find a function $x(t)$ that is essentially the same as its derivative $\dot{x}$ times a constant λ. The family of functions with this property are the exponentials and if we try

$$x(t) = e^{\lambda t} \quad \text{we find} \quad \dot{x}(t) = \lambda e^{\lambda t} \quad \text{hence} \quad \dot{x} = \lambda x \qquad (3)$$

and therefore $x(t)$ is a solution. In fact if we multiply the exponential by a constant c it also satisfies (2)

$$x(t) = c e^{\lambda t} \quad \text{we find} \quad \dot{x}(t) = c \lambda e^{\lambda t} \quad \text{and still} \quad \dot{x} = \lambda x \qquad (4)$$

But now these are infinitely many functions – we have found the general solution of (2) – and we leave it to the mathematicians to prove that these are the only functions that fulfill (2) and that we have found all of them, i.e. *uniqueness* and *completeness* of the solutions. It turns out that the general solution of a dynamical system of nth order has n open constants and as we are dealing with one-dimension systems here we have one open constant: the c in the above solution. The constant c can be determined if we know the state of the system at a given time t, for instance $x(t = 0) = x_0$

$$x(t = 0) = x_0 = c e^0 \quad \rightarrow \quad c = x_0 \qquad (5)$$

where x_0 is called the *initial condition*. Figure 1 shows plots of the solutions of (2) for different initial conditions and parameter values $\lambda < 0$, $\lambda = 0$ and $\lambda > 0$.

We now turn to the question whether it is possible to get an idea of the dynamical properties of (2) or (1) *without* calculating solutions, which, as mentioned above, is not possible in general anyway. We start with (2) as we know the solution in this

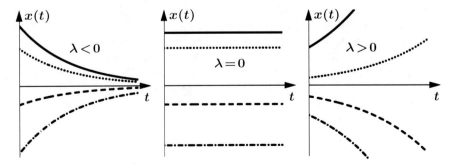

Fig. 1 Solutions $x(t)$ for the equation of continuous growth (2) for different initial conditions x_0 (solid, dashed, dotted and dash-dotted) and parameter values $\lambda < 0$, $\lambda = 0$ and $\lambda > 0$ on the left, in the middle and on the right, respectively.

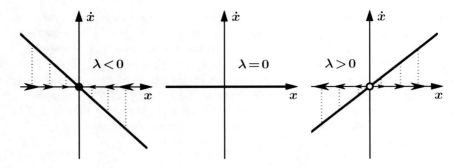

Fig. 2 Phase space plots, $\dot{x}$ as a function of x, for the equation of continuous growth (2) for the cases $\lambda < 0$, $\lambda = 0$ and $\lambda > 0$ on the left, in the middle and on the right, respectively.

case and now plot $\dot{x}$ as a function of x, a representation called a *phase space plot* and shown in fig. 2, again for $\lambda < 0$, $\lambda = 0$ and $\lambda > 0$. The graphs are straight lines given by $\dot{x} = \lambda x$ with a negative, vanishing and positive slope, respectively. So what can we learn from these graphs? The easiest is the one in the middle corresponding to $\dot{x} = 0$, which means there are no changes in the system. Where ever we start initially we stay there, a quite boring case.

Next we turn to the plot on the left, $\lambda < 0$, for which the phase space plot is a straight line with a negative slope. So for any state $x < 0$ the change $\dot{x}$ is positive, the system evolves to the right. Moreover, the more negative the state x the bigger the change $\dot{x}$ towards the origin as indicated by the direction and size of the arrows on the horizontal axis. In contrast, for any initial positive state $x > 0$ the change $\dot{x}$ is negative and the system evolves towards the left. In both cases it is approaching the origin and the closer it gets the more it slows down. For the system (2) with $\lambda < 0$ all *trajectories* evolve towards the origin, which is therefore called a *stable fixed point* or *attractor*. Fixed points and their stability are most important properties of dynamical systems, in particular for nonlinear systems as we shall see later. In phase space plots like fig. 2 stable fixed points are indicated by solid circles.

On the right in fig. 2 the case for $\lambda > 0$ is depicted. Here, for any positive (negative) state x the change $\dot{x}$ is also positive (negative) as indicated by the arrows and the system moves away from the origin in both direction. Therefore, the origin in this case is an *unstable fixed point* or *repeller* and indicated by an open circle in the phase space plot. Finally, coming back to $\lambda = 0$ shown in the middle of fig. 2, all points on the horizontal axis are fixed points. However, they are neither attracting nor repelling and are therefore called *neutrally stable*.

1.2 Nonlinear Systems: First Steps

The concepts discussed in the previous section for the linear equation of continuous growth can immediately be applied to nonlinear systems in one dimension. To be most explicit we treat an example known as the *logistic equation*

$$\dot{x} = \lambda x - x^2 \qquad (6)$$

The graph of this function is a parabola which opens downwards, it has one intersection with the horizontal axis at the origin and another one at $x = \lambda$ as shown in fig. 3.

These intersections between the graph and the horizontal axis are most important because they are the fixed points of the system, i.e. the values of x for which $\dot{x} = 0$ is fulfilled. For the case $\lambda < 0$, shown on the left in fig. 3, the graph intersects the negative x-axis with a positive slope. As we have seen above – and of course one can apply the reasoning regarding the state and its change here again – such a slope means that the system is moving away from this point, which is therefore classified as an unstable fixed point or repeller. The opposite is case for the fixed point at the origin. The flow moves towards this location from both side, so it is stable or an attractor. Corresponding arguments can be made for $\lambda > 0$ shown on the right in fig. 3.

An interesting case is $\lambda = 0$ shown in the middle of fig. 3. Here the slope vanishes, a case we previously called neutrally stable. However, by inspecting the state and change in the vicinity of the origin, it is easily determined that the flow moves towards this location if we are on the positive x-axis and away from it when x is negative. Such points are called *half-stable* or *saddle points* and denoted by half-filled circles.

As a second example we discuss the cubic equation

$$\dot{x} = \lambda x - x^3 \qquad (7)$$

From the graph of this function, shown in fig. 4, it is evident that for $\lambda \leq 0$ there is one stable fixed point at the origin which becomes unstable when λ is increased to positive values and at the same time two stable fixed points appear to its right and left. Such a situation, where more than one stable state exist in a system is called *multistability*, in the present case of two stable fixed points *bistability*, an inherently nonlinear property which does not exist in linear systems. Moreover, (7)

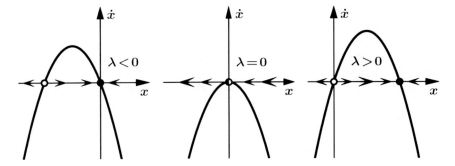

Fig. 3 Phase space plots, $\dot{x}$ as a function of x, for the logistic equation (6) for the cases $\lambda < 0$, $\lambda = 0$ and $\lambda > 0$ on the left, in the middle and on the right, respectively.

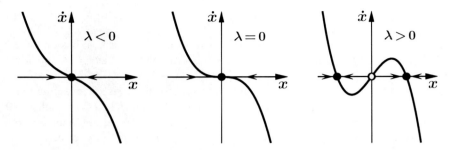

Fig. 4 Phase space plots, $\dot{x}$ as a function of x, for the cubic equation (7) for the cases $\lambda < 0$, $\lambda = 0$ and $\lambda > 0$ on the left, in the middle and on the right, respectively.

becomes bistable when the parameter λ switches from negative to positive values. When this happens, the change in the system's dynamical behavior is not gradual but *qualitative*. A system, which was formerly *monostable* with a single attractor at the origin, now has become *bistable* with three fixed points, two of them stable and the origin having switched from an attractor to a repeller. It is this kind of qualitative change in behavior when a parameter exceeds a certain threshold that makes nonlinear differential equations the favorite modeling tool to describe the transition phenomena we observe in nature.

1.3 Potential Functions

So far we derived the dynamical properties of linear and nonlinear systems from their phase space plots. There is another, arguably even more intuitive way to find out about a system's behavior, which is by means of potential functions. In one-dimensional systems the potential is defined by

$$\dot{x} = f(x) = -\frac{dV}{dx} \quad \rightarrow \quad V(x) = -\int f(x)\,dx + c \qquad (8)$$

In words: the negative derivative of the potential function is the right-hand side of the differential equation. All one-dimensional systems have a potential, even though it may not be possible to write it down in a closed analytical form. For higher dimensional systems the existence of a potential is more the exception than the rule as we shall see in sect. 2.5.

From its definition (8) it is obvious that the change in state $\dot{x}$ is equal to the negative slope of the potential function. First, this implies that the system always moves in the direction where the potential is decreasing and second, that the fixed points of the system are located at the extrema of the potential, where minima correspond to stable and maxima to unstable fixed points. The dynamics of a system can be seen as the overdamped motion of a particle the landscape of the potential. One can think of an overdamped motion as the movement of a particle in a thick or viscous fluid

Dynamical Systems in One and Two Dimensions

like honey. If it reaches a minimum it will stick there, it will not oscillate back and forth.

Examples

1. $\dot{x} = \lambda x = -\dfrac{dV}{dx} \quad \rightarrow \quad V(x) = -\int \lambda x \, dx + c = -\dfrac{1}{2}\lambda x^2 \underbrace{+c}_{=0}$

   The familiar linear equation. Plots of $\dot{x}$ and the corresponding potential V as functions of x are shown in fig. 5 for the cases $\lambda < 0$ (left) and $\lambda > 0$ (middle);

2. $\dot{x} = x - x^2 = -\dfrac{dV}{dx} \quad \rightarrow \quad V(x) = -\dfrac{1}{2}x^2 + \dfrac{1}{3}x^3$

   A special case of the logistic equation. The potential in this case is a cubic function shown in fig. 5 (right);

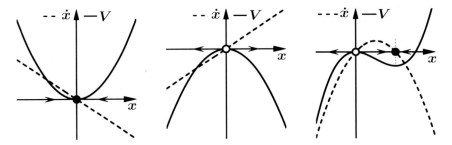

Fig. 5 Graphs of $\dot{x}$ (dashed) and $V(x)$ (solid) for the linear equation ($\lambda < 0$ left, $\lambda > 0$ middle) and for the logistic equation (right).

3. $\dot{x} = \lambda x - x^3 \quad \rightarrow \quad V(x) = -\dfrac{1}{2}\lambda x^2 + \dfrac{1}{4}x^4$

   The cubic equation for which graphs and potential functions are shown in fig. 6. Depending on the sign of the parameter λ this system has either a single attractor at the origin or a pair of stable fixed points and one repeller.

4. $\dot{x} = \lambda + x - x^3 \quad \rightarrow \quad V(x) = -\lambda x - \dfrac{1}{2}x^2 + \dfrac{1}{4}x^4$

   For the case $\lambda = 0$ this equation is a special case of the cubic equation we have dealt with above, namely $\dot{x} = x - x^3$. The phase space plots for this special case are shown in fig. 7 in the left column. The top row in this figure shows what is happening when we increase λ from zero to positive values. We are simply adding a constant, so the graph gets shifted upwards. Correspondingly, when we decrease λ from zero to negative values the graph gets shifted downwards, as shown in the bottom row in fig. 7.

   The important point in this context is the number of intersections of the graphs with the horizontal axis, i.e. the number of fixed points. The special case with $\lambda = 0$ has three as we know and if we increase or decrease λ only slightly this number stays the same. However, there are certain values of λ, for which one of

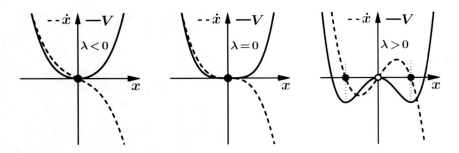

Fig. 6 Graph of $\dot{x}$ (dashed) and $V(x)$ (solid) for the cubic equation for different values of λ.

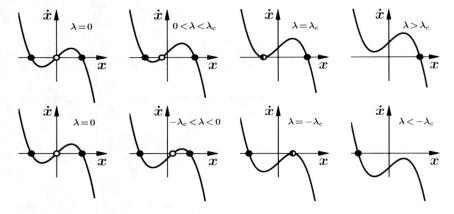

Fig. 7 Phase space plots for $\dot{x} = \lambda + x - x^3$. For positive (negative) values of λ the graphs are shifted up (down) with respect to the point symmetric case $\lambda = 0$ (left column). The fixed point skeleton changes at the critical parameter values $\pm \lambda_c$.

the extrema is located on the horizontal axis and the system has only two fixed points as can be seen in the third column in fig. 7. We call these the critical values for the parameter, $\pm \lambda_c$. A further increase or decrease beyond these critical values leaves the system with only one fixed point as shown in the rightmost column. Obviously, a qualitative change in the system occurs at the parameter values $\pm \lambda_c$ when a transition from three fixed points to one fixed point takes place.

A plot of the potential functions where the parameter is varied from $\lambda < -\lambda_c$ to $\lambda > \lambda_c$ is shown in fig. 8. In the graph on the top left for $\lambda < -\lambda_c$ the potential has a single minimum corresponding to a stable fixed point, as indicated by the gray ball, and the trajectories from all initial conditions end there. If λ is increased a half-stable fixed point emerges at $\lambda = -\lambda_c$ and splits into a stable and unstable fixed point, i.e. a local minimum and maximum when the parameter exceeds this threshold. However, there is still the local minimum for negative values of x and the system, represented by the gray ball, will remain there. It takes an increase in λ beyond λ_c in the bottom row before this minimum disappears and the system switches to the only remaining fixed point on the right. Most importantly, the dynamical behavior is different if we start with a $\lambda > \lambda_c$, as in the graph at the bottom right and decrease the control parameter. Now the gray ball will stay at positive values of x until the critical value $-\lambda_c$ is passed and the system switches to the left. The state of the system does not only depend on the value of the control parameter but also on its history of parameter changes – it has a form of memory. This important and wide spread phenomenon is called *hysteresis* and we shall come back to it in sect. 1.4.

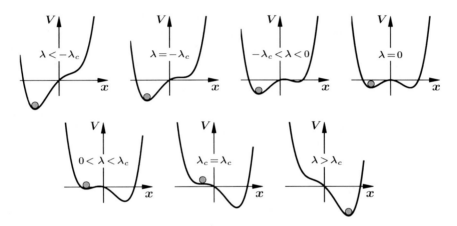

Fig. 8 Potential functions for $\dot{x} = \lambda + x - x^3$ for parameter values $\lambda < -\lambda_c$ (top left) to $\lambda > \lambda_c$ (bottom right). If a system, indicated by the gray ball, is initially in the left minimum, λ has to increase beyond λ_c before a switch to the right minimum takes place. In contrast, if the system is initially in the right minimum, λ has to decrease beyond $-\lambda_c$ before a switch occurs. The system shows hysteresis.

1.4 Bifurcation Types

One major difference between linear and nonlinear systems is that the latter can undergo qualitative changes when a parameter exceeds a critical value. So far we have characterized the properties of dynamical systems by phase space plots and potential functions for different values of the control parameter, but it is also possible to display the locations and stability of fixed points as a function of the parameter in a single plot, called a *bifurcation diagram*. In these diagrams the locations of stable fixed points are represented by solid lines, unstable fixed points are shown dashed. We shall also use solid, open and half-filled circles to mark stable, unstable and half-stable fixed points, respectively.

There is a quite limited number of ways how such qualitative changes, also called *bifurcations*, can take place in one-dimensional systems. In fact, there are four basic types of bifurcations known as *saddle-node, transcritical*, and *super-* and *subcritical pitchfork bifurcation*, which we shall discuss. For each type we are going to show a plot with the graphs in phase space at the top, the potentials in the bottom row, and in-between the bifurcation diagram with the fixed point locations $\tilde{x}$ as functions of the control parameter λ.

Saddle-Node Bifurcation

The prototype of a system that undergoes a saddle-node bifurcation is given by

$$\dot{x} = \lambda + x^2 \quad \rightarrow \quad \tilde{x}_{1,2} = \pm\sqrt{-\lambda} \tag{9}$$

The graph in phase space for (9) is a parabola that open upwards. For negative values of λ one stable and one unstable fixed point exist, which collide and annihilate when λ is increased above zero. There are no fixed points in this system for positive values of λ. Phase space plots, potentials and a bifurcation diagram for (9) are shown in fig. 9.

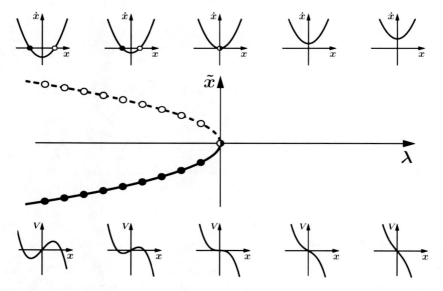

Fig. 9 Saddle-node bifurcation: a stable and unstable fixed point collide and annihilate. Top: phase space plots; middle: bifurcation diagram; bottom: potential functions.

Transcritical Bifurcation

The transcritical bifurcation is given by

$$\dot{x} = \lambda x + x^2 \quad \rightarrow \quad \tilde{x}_1 = 0, \ \tilde{x}_2 = \lambda \tag{10}$$

and summarized in fig. 10. For all parameter values, except the bifurcation point $\lambda = 0$, the system has a stable and an unstable fixed point. The bifurcation diagram

Dynamical Systems in One and Two Dimensions 11

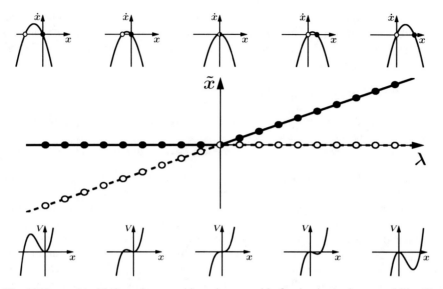

Fig. 10 Transcritical bifurcation: a stable and an unstable fixed point exchange stability. Top: phase space plots; middle: bifurcation diagram; bottom: potential functions.

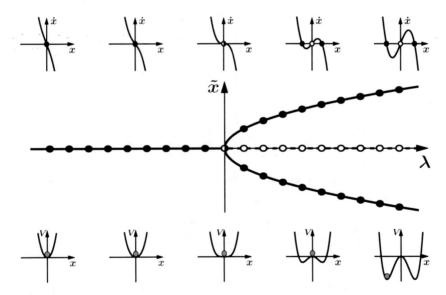

Fig. 11 Supercritical pitchfork bifurcation: a stable fixed point becomes unstable and two new stable fixed points arise. Top: phase space plots; middle: bifurcation diagram; bottom: potential functions.

consists of two straight lines, one at $\tilde{x} = 0$ and one with a slope of one. When these lines intersect at the origin they exchange stability, i.e. former stable fixed points along the horizontal line become unstable and the repellers along the line with slope one become attractors.

Supercritical Pitchfork Bifurcation

The supercritical pitchfork bifurcation is visualized in fig. 11 and is prototypically given by

$$\dot{x} = \lambda x - x^3 \quad \to \quad \tilde{x}_1 = 0, \; \tilde{x}_{2,3} = \pm\sqrt{\lambda} \tag{11}$$

The supercritical pitchfork bifurcation is the main mechanism for switches between mono- and bistability in nonlinear systems. A single stable fixed point at the origin becomes unstable and a pair of stable fixed points appears symmetrically around $\tilde{x} = 0$. In terms of symmetry this system has an interesting property: the differential equation (11) is invariant if we substitute x by $-x$. This can also be seen in the phase space plots, which all have a point symmetry with respect to the origin, and in the plots of the potential, which have a mirror symmetry with respect to the vertical axis. If we prepare the system with a parameter $\lambda < 0$ it will settle down at the only fixed point, the minimum of the potential at $x = 0$, as indicated by the gray ball in fig. 11 (bottom left). The potential together with the solution still have the mirror symmetry with respect to the vertical axis. If we now increase the parameter beyond its critical value $\lambda = 0$, the origin becomes unstable as can be seen in fig. 11 (bottom second

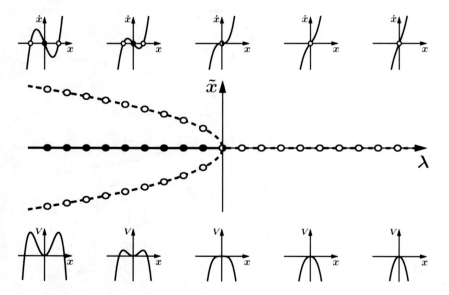

Fig. 12 Subcritical pitchfork bifurcation: a stable and two unstable fixed points collide and the former attractor becomes a repeller. Top: phase space plots; middle: bifurcation diagram; bottom: potential functions.

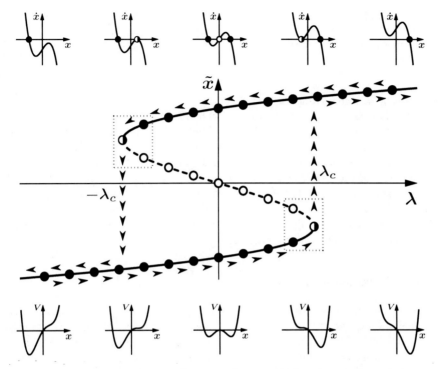

Fig. 13 A system showing hysteresis. Depending on whether the parameter is increased from large negative or decreased from large positive values the switch occurs at $\lambda = \lambda_c$ or $\lambda = -\lambda_c$, respectively. The bifurcation is not a basic type but consists of two saddle-node bifurcations indicated by the dotted rectangles. Top: phase space plots; middle: bifurcation diagram; bottom: potential functions.

from right). Now the slightest perturbation will move the ball to the left or right where the slope is finite and it will settle down in one of the new minima (fig. 11 (bottom right)). At this point, the potential plus solution is not symmetric anymore, the symmetry of the system has been broken by the solution. This phenomenon, called *spontaneous symmetry breaking*, is found in many systems in nature.

Subcritical Pitchfork Bifurcation

The equation governing the subcritical pitchfork bifurcation is given by

$$\dot{x} = \lambda x + x^3 \quad \rightarrow \quad \tilde{x}_1 = 0, \; \tilde{x}_{2,3} = \pm\sqrt{-\lambda} \tag{12}$$

and and its diagrams are shown in fig. 12. As in the supercritical case the origin is stable for negative values of λ and becomes unstable when the parameter exceeds $\lambda = 0$. Two additional fixed points exist for negative parameter values at $\tilde{x} = \pm\sqrt{-\lambda}$ and they are repellers.

System with Hysteresis

As we have seen before the system

$$\dot{x} = \lambda + x - x^3 \tag{13}$$

shows hysteresis, a phenomenon best visualized in the bifurcation diagram in fig. 13. If we start at a parameter value below the critical value $-\lambda_c$ and increase λ slowly, we will follow a path indicated by the arrows below the lower solid branch of stable fixed points in the bifurcation diagram. When we reach $\lambda = \lambda_c$ this branch does not continue and the system has to jump to the upper branch. Similarly, if we start at a large positive value of λ and decrease the parameter, we will stay on the upper branch of stable fixed points until we reach the point $-\lambda_c$ from where there is no smooth way out and a discontinuous switch to the lower branch occurs.

It is important to realize that (13) is not a basic bifurcation type. In fact, it consists of two saddle-node bifurcations indicated by the dotted rectangles in fig. 13.

2 Two-Dimensional Systems

Two-dimensional dynamical systems can be represented by either a single differential equation of second order, which contains a second derivative with respect to time, or by two equations of first order. In general, a second order system can always be expressed as two first order equations, but most first order systems cannot be written as a single second order equation

$$\ddot{x} + f(x, \dot{x}) = 0 \quad \rightarrow \quad \begin{cases} \dot{x} = y \\ \dot{y} = -f(x, y = \dot{x}) \end{cases} \tag{14}$$

2.1 Linear Systems and their Classification

A general linear two-dimensional system is given by

$$\begin{aligned} \dot{x} &= ax + by \\ \dot{y} &= cx + dy \end{aligned} \tag{15}$$

and has a fixed at the origin $\tilde{x} = 0$, $\tilde{y} = 0$.

The Pedestrian Approach

One may ask the question whether it is possible to decouple this system somehow, such that $\dot{x}$ only depends on x and $\dot{y}$ only on y. This would mean that we have two one-dimensional equations instead of a two-dimensional system. So we try

$$\begin{aligned} \dot{x} &= \lambda x \\ \dot{y} &= \lambda y \end{aligned} \quad \rightarrow \quad \begin{aligned} ax + by &= \lambda x \\ cx + dy &= \lambda y \end{aligned} \quad \rightarrow \quad \begin{aligned} (a-\lambda)x + by &= 0 \\ cx + (d-\lambda)y &= 0 \end{aligned} \tag{16}$$

Dynamical Systems in One and Two Dimensions

where we have used (15) and obtained a system of equations for x and y. Now we are trying to solve this system

$$y = -\frac{a-\lambda}{b}x \quad \to \quad cx - \frac{(a-\lambda)(d-\lambda)}{b}x = 0$$
$$\to \quad [\underbrace{(a-\lambda)(d-\lambda) - bc}_{=0}]x = 0 \tag{17}$$

From the last term it follows that $x = 0$ is a solution, in which case form the first equation follows $y = 0$. However, there is obviously a second way how this system of equation can be solved, namely, if the under-braced term inside the brackets vanishes. Moreover, this term contains the parameter λ, which we have introduced in a kind of ad hoc fashion above, and now can be determined such that this term actually vanishes

$$(a-\lambda)(d-\lambda) - bc = 0 \quad \to \quad \lambda^2 - (a+d)\lambda + ad - bc = 0$$
$$\to \quad \lambda_{1,2} = \tfrac{1}{2}\{a + d \pm \sqrt{(a+d)^2 - 4(ad-bc)}\} \tag{18}$$

For simplicity, we assume $a = d$, which leads to

$$\lambda_{1,2} = a \pm \frac{1}{2}\sqrt{4a^2 - 4a^2 + 4bc} = a \pm \sqrt{bc} \tag{19}$$

As we know λ now, we can go back to the first equation in (17) and calculate y

$$y = -\frac{a-\lambda}{b}x = -\frac{a - (a \pm \sqrt{bc})}{b}x = \pm\sqrt{\frac{c}{b}}x \tag{20}$$

So far, so good but we need to figure out what this all means. In the first step we assumed $\dot{x} = \lambda x$ and $\dot{y} = \lambda y$. As we know from the one-dimensional case, such systems are stable for $\lambda < 0$ and unstable for $\lambda > 0$. During the calculations above we found two possible values for lambda, $\lambda_{1,2} = a \pm \sqrt{bc}$, which depend on the parameters of the dynamical system $a = d$, b and c. Either of them can be positive or negative, in fact if the product bc is negative, the λs can even be complex. For now we are going to exclude the latter case, we shall deal with it later. In addition, we have also found a relation between x and y for each of the values of λ, which is given by (20). If we plot y as a function of x (20) defines two straight lines through the origin with slopes of $\pm\sqrt{c/b}$, each of these lines corresponds to one of the values of lambda and the dynamics along these lines is given by $\dot{x} = \lambda x$ and $\dot{y} = \lambda y$. Along each of these lines the system can either approach the origin from both sides, in which cases it is called a stable direction or move away from it, which means the direction is unstable. Moreover, these are the only directions in the xy-plane where the dynamics evolves along straight lines and therefore built up a skeleton from which other trajectories can be easily constructed. Mathematically, the λs are called the eigenvalues and the directions represent the eigenvectors of the coefficient matrix as we shall see next.

There are two important descriptors of a matrix in this context, the trace and the determinant. The former is given by the sum of the diagonal elements $t_r = a + d$ and the latter, for a 2×2 matrix, is the difference between the products of the upper-left times lower-right and upper-right times lower-left elements $d_{et} = ad - bc$.

The Matrix Approach

Any two-dimensional linear system can be written in matrix form

$$\dot{\mathbf{x}} = \begin{pmatrix} a & b \\ c & d \end{pmatrix} \mathbf{x} = L\mathbf{x} \quad \rightarrow \quad \dot{\tilde{\mathbf{x}}} = \begin{pmatrix} 0 \\ 0 \end{pmatrix} \tag{21}$$

with a fixed point at the origin. If a linear system's fixed point is not at the origin a coordinate transformation can be applied that shifts the fixed point such that (21) is fulfilled. The eigenvalues of L can be readily calculated and it is most convenient to express them in terms of the trace and determinant of L

$$\begin{vmatrix} a-\lambda & b \\ c & d-\lambda \end{vmatrix} = \lambda^2 - \lambda \underbrace{(a+d)}_{\text{trace } t_r} + \underbrace{ad-bc}_{\text{determinant } d_{et}} = 0 \tag{22}$$

$$\begin{aligned} \rightarrow \quad \lambda_{1,2} &= \tfrac{1}{2}\{a+d \pm \sqrt{(a+d)^2 - 4(ad-bc)}\} \\ &= \tfrac{1}{2}\{t_r \pm \sqrt{t_r^2 - 4d_{et}}\} \end{aligned} \tag{23}$$

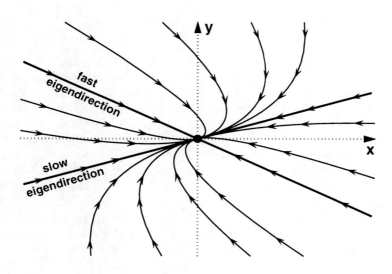

Fig. 14 Phase space portrait for the stable node.

Depending on whether the discriminant $t_r^2 - 4d_{et}$ in (23) is bigger or smaller than zero, the eigenvalues $\lambda_{1,2}$ will be real or complex numbers, respectively.

$$t_r^2 - 4d_{et} > 0 \quad \rightarrow \quad \lambda_{1,2} \in \mathbb{R}$$

If both eigenvalues are negative, the origin is a stable fixed point, in this case called a *stable node*. An example of trajectories in the two-dimensional phase space is shown in fig. 14. We assume the two eigenvalues to be unequal, $\lambda_1 < \lambda_2$ and both smaller than zero. Then, the only straight trajectories are along the eigendirections which are given by the eigenvectors of the system. All other trajectories are curved as the rate of convergence is different for the two eigendirections depending on the corresponding eigenvalues. As we assumed $\lambda_1 < \lambda_2$ the trajectories approach the fixed point faster along the direction of the eigenvector $\mathbf{v}^{(1)}$ which corresponds to λ_1 and is therefore called the *fast eigendirection*. In the same way, the direction related to λ_2 is called the *slow eigendirection*.

Correspondingly, for the phase space plot when both eigenvalues are positive the flow, as indicated by the arrows in fig. 14, is reversed and leads away from the fixed point which is then called an *unstable node*.

For the degenerate case, with $\lambda_1 = \lambda_2$ we have a look at the system with

$$L = \begin{pmatrix} -1 & b \\ 0 & -1 \end{pmatrix} \quad \rightarrow \quad \lambda_{1,2} = -1 \qquad (24)$$

The eigenvectors are given by

$$\begin{pmatrix} -1 & b \\ 0 & -1 \end{pmatrix} = \begin{pmatrix} v_1 \\ v_2 \end{pmatrix} \quad \rightarrow \quad \begin{matrix} -v_1 + bv_2 = -v_1 \\ -v_2 = -v_2 \end{matrix} \quad \rightarrow \quad bv_2 = 0 \qquad (25)$$

For $b \neq 0$ the only eigendirection of L is the horizontal axis with $v_2 = 0$. The fixed point is called a *degenerate node* and its phase portrait shown in fig. 15 (left). If $b = 0$ any vector is an eigenvector and the trajectories are straight lines pointing towards or away from the fixed point depending on the sign of the eigenvalues. The phase space portrait for this situation is shown in fig. 15 (right) and the fixed point is for obvious reasons called a *star node*.

If one of the eigenvalues is positive and the other negative, the fixed point at the origin is half-stable and called a *saddle point*. The eigenvectors define the directions where the flow in phase space is pointing towards the fixed point, the so-called *stable direction*, corresponding to the negative eigenvalue, and away from the fixed point, the *unstable direction*, for the eigenvector with a positive eigenvalue. A typical phase space portrait for a saddle point is shown in fig. 16.

$$t_r^2 - 4d_{et} < 0 \quad \rightarrow \quad \lambda_{1,2} \in \mathbb{C} \quad \rightarrow \quad \lambda_2 = \lambda_1^*$$

If the discriminant $t_r^2 - 4d_{et}$ in (23) is negative the linear two-dimensional system has a pair of complex conjugate eigenvalues. The stability of the fixed point is then

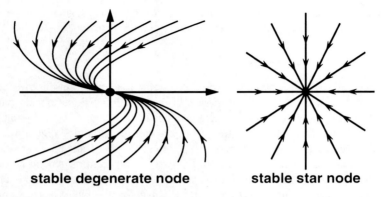

Fig. 15 Degenerate case where the eigenvalues are the same. The degenerate node (left) has only one eigendirection, the star node (right) has infinitely many.

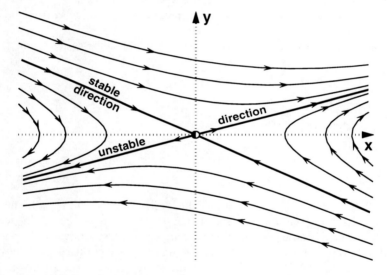

Fig. 16 If the eigenvalues have different signs $\lambda_1 \lambda_2 < 0$ the fixed point at the origin is half-stable and called a saddle point.

determined by the real part of the eigenvalues given as the trace of the coefficient matrix L in (21). The trajectories in phase space are spiraling towards or away from the origin as a *stable spiral* for a negative real part of the eigenvalue or an *unstable spiral* if the real part is positive as shown in fig. 17 left and middle, respectively. A special case exists when the real part of the eigenvalues vanishes $t_r = 0$. As can be seen in fig. 17 (right) the trajectories are closed orbits. The fixed point at the origin is neutrally stable and called a *center*.

Dynamical Systems in One and Two Dimensions

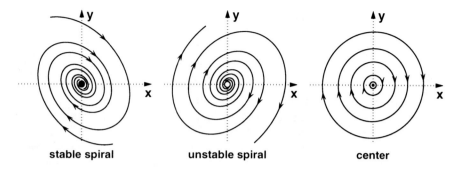

Fig. 17 For complex eigenvalues the trajectories in phase space are stable spirals if their real part is negative (left) and unstable spirals for a positive real part (middle). If the real part of the eigenvalues vanishes the trajectories are closed orbits around the origin, which is then a neutrally stable fixed point called a center (right).

To summarize these findings, we can now draw a diagram in a plane as shown in fig. 18, where the axes are the determinant d_{et} and trace t_r of the linear matrix L that provides us with a complete classification of the linear dynamical systems in two dimensions.

On the left of the vertical axis ($d_{et} < 0$) are the saddle points. On the right ($d_{et} > 0$) are the centers on the horizontal axis ($t_r = 0$) with unstable and stable spirals located above and below, respectively. The stars and degenerate nodes are along the parabola $t_r^2 = 4d_{et}$ that separates the spirals from the stable and unstable nodes.

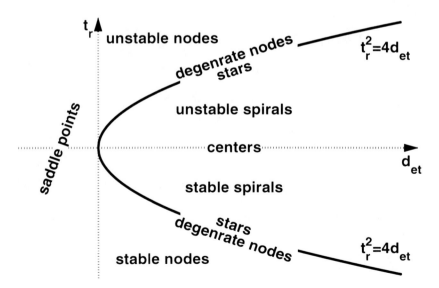

Fig. 18 Classification diagram for two-dimensional linear systems in terms of the trace t_r and determinant d_{et} of the linear matrix.

2.2 Nonlinear Systems

In general a two dimensional dynamical system is given by

$$\dot{x} = f(x,y) \qquad \dot{y} = g(x,y) \tag{26}$$

In the one-dimensional systems the fixed points were given by the intersection between the function in phase space and the horizontal axis. In two dimensions we can have graphs that define the locations $\dot{x} = f(x,y) = 0$ and $\dot{y} = g(x,y) = 0$, which are called the *nullclines* and are the location in the xy-plane where the tangent to the trajectories is vertical or horizontal, respectively. Fixed points are located at the intersections of the nullclines. We have also seen in one-dimensional systems that the stability of the fixed points is given by the slope of the function at the fixed point. For the two-dimensional case the stability is also related to derivatives but now there is more than one, there is the so-called *Jacobian matrix*, which has to be evaluated at the fixed points

$$J = \begin{pmatrix} \frac{\partial f}{\partial x} & \frac{\partial f}{\partial y} \\ \frac{\partial g}{\partial x} & \frac{\partial g}{\partial y} \end{pmatrix} \tag{27}$$

The eigenvalues and eigenvectors of this matrix determine the nature of the fixed points, whether it is a node, star, saddle, spiral or center and also the dynamics in its vicinity, which is best shown in a detailed example.

Detailed Example

We consider the two-dimensional system

$$\dot{x} = f(x,y) = y - y^3 = y(1-y^2), \quad \dot{y} = g(x,y) = -x - y^2 \tag{28}$$

for which the nullclines are given by

$$\begin{aligned} \dot{x} = 0 &\rightarrow y = 0 \text{ and } y = \pm 1 \\ \dot{y} = 0 &\rightarrow y = \pm\sqrt{-x} \end{aligned} \tag{29}$$

The fixed points are located at the intersections of the nullclines

$$\tilde{\mathbf{x}}_1 = \begin{pmatrix} 0 \\ 0 \end{pmatrix} \qquad \tilde{\mathbf{x}}_{2,3} = \begin{pmatrix} -1 \\ \pm 1 \end{pmatrix} \tag{30}$$

We determine the Jacobian of the system by calculating the partial derivatives

$$J = \begin{pmatrix} \frac{\partial f}{\partial x} & \frac{\partial f}{\partial y} \\ \frac{\partial g}{\partial x} & \frac{\partial g}{\partial y} \end{pmatrix} = \begin{pmatrix} 0 & 1 - 3y^2 \\ -1 & -2y \end{pmatrix} \tag{31}$$

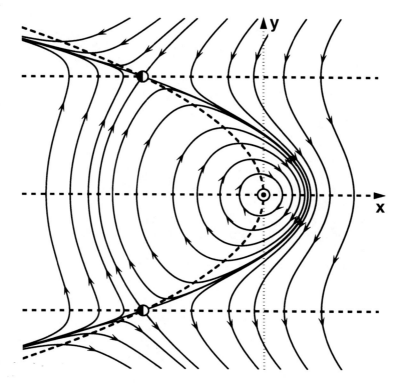

Fig. 19 Phase space diagram for the system (28).

A phase space plot for the system (28) is shown in fig. 19. The origin is a center surrounded by closed orbits with flow in the clockwise direction. This direction is readily determined by calculating the derivatives close to the origin

$$\dot{\mathbf{x}} = \begin{pmatrix} y - y^3 \\ -x - y^2 \end{pmatrix} \quad \text{at} \quad \mathbf{x} = \begin{pmatrix} 0.1 \\ 0 \end{pmatrix} \quad \rightarrow \quad \dot{\mathbf{x}} = \begin{pmatrix} 0 \\ -0.1 \end{pmatrix} \quad \rightarrow \quad \text{clockwise}$$

The slope of the trajectories at the two saddle points is given by the direction of their eigenvectors, and whether a particular direction is stable or unstable is determined by the corresponding eigenvalues. The two saddles are connected by two trajectories and such connecting trajectories between two fixed points are called *heteroclinic orbits*. The dashed horizontal lines through the fixed points and the dashed parabola which opens to the left are the nullclines where the the trajectories are either vertical or horizontal.

Second Example: Homoclinic Orbit

In the previous example we encountered heteroclinic orbits, which are trajectories that leave a fixed point along one of its unstable directions and approach another fixed point along a stable direction. In a similar way it is also possible that the trajectory returns along a stable direction to the fixed point it originated from. Such a closed trajectory that starts and ends at the same fixed point is correspondingly called a *homoclinic orbit*. To be specific we consider the system

$$\dot{x} = y - y^2 = y(1-y) \qquad \rightarrow \qquad \tilde{\mathbf{x}}_1 = \begin{pmatrix} 0 \\ 0 \end{pmatrix} \qquad \tilde{\mathbf{x}}_2 = \begin{pmatrix} 0 \\ 1 \end{pmatrix} \tag{32}$$

with the Jacobian matrix

$$J = \begin{pmatrix} 0 & 1-2y \\ 1 & 0 \end{pmatrix} \qquad \rightarrow \qquad J(\tilde{\mathbf{x}}_{1,2}) = \begin{pmatrix} 0 & \pm 1 \\ 1 & 0 \end{pmatrix} \tag{33}$$

From $t_r[J(\tilde{\mathbf{x}}_1)] = 0$ and $d_{et}[J(\tilde{\mathbf{x}}_1)] = -1$ we identify the origin as a saddle point. In the same way with $t_r[J(\tilde{\mathbf{x}}_2)] = 0$ and $d_{et}[J(\tilde{\mathbf{x}}_2)] = 1$ the second fixed point is classified as a center.

The eigenvalues and eigenvectors are readily calculated

$$\tilde{\mathbf{x}}_1: \ \lambda_{1,2}^{(1)} = \pm\sqrt{2}, \quad \mathbf{v}_{1,2}^{(1)} = \begin{pmatrix} 1 \\ \pm\sqrt{2} \end{pmatrix} \qquad \tilde{\mathbf{x}}_2: \ \lambda_{1,2}^{(2)} = \pm i\sqrt{2} \tag{34}$$

The nullclines are given by $y = 0$, $y = 1$ (vertical) and $x = 0$ (horizontal).

A phase space plot for the system (32) is shown in fig. 20 where the fixed point at the origin has a homoclinic orbit. The trajectory is leaving $\tilde{\mathbf{x}}_1$ along the unstable direction, turning around the center $\tilde{\mathbf{x}}_2$ and returning along the stable direction of the saddle.

2.3 Limit Cycles

A limit cycle, the two-dimensional analogon of a fixed point, is an *isolated closed* trajectory. Consequently, limit cycles exist with the flavors *stable*, *unstable* and *half-stable* as shown in fig. 21. A stable limit cycle attracts trajectories from both its outside and its inside, whereas an unstable limit cycle repels trajectories on both sides. There also exist closed trajectories, called half-stable limit cycles, which attract the trajectories from one side and repel those on the other. Limit cycles are inherently nonlinear objects and must not be mixed up with the centers found in the previous section in linear systems when the real parts of both eigenvalues vanish. These centers are not isolated closed trajectories, in fact there is always another closed trajectory infinitely close nearby. Also all centers are neutrally stable, they are neither attracting nor repelling.

From fig. 21 it is intuitively clear that inside a stable limit cycle, there must be an unstable fixed point or an unstable limit cycle, and inside an unstable limit cycle there is a stable fixed point or a stable limit cycle. In fact, this intuition will guide us to a new and one of the most important bifurcation types: the *Hopf bifurcation*.

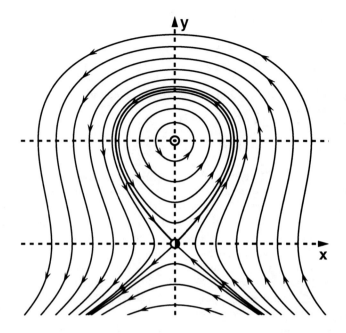

Fig. 20 Phase space diagram with a homoclinic orbit.

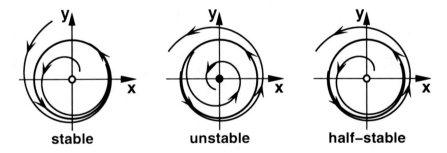

Fig. 21 Limit cycles attracting or/and repelling neighboring trajectories.

2.4 Hopf Bifurcation

We consider the dynamical system

$$\dot{\xi} = \mu\, \xi - \xi\, |\xi|^2 \quad \text{with} \quad \mu, \xi \in \mathbb{C} \tag{35}$$

where both the parameter μ and the variable ξ are complex numbers. There are essentially two ways in which complex numbers can be represented

1. Cartesian representation: $\xi = x + iy$ for which (35) takes the form

$$\dot{x} = \varepsilon x - \omega y - x(x^2 + y^2)$$
$$\dot{y} = \varepsilon x + \omega y - y(x^2 + y^2) \quad (36)$$

after assuming $\mu = \varepsilon + i\omega$ and splitting into real and imaginary part;

2. Polar representation: $\xi = re^{i\varphi}$ and (35) becomes

$$\dot{r} = \varepsilon r - r^3 \qquad \dot{\varphi} = \omega \quad (37)$$

Rewriting (35) in a polar representation leads to a separation of the complex equation not into a coupled system as in the cartesian case (36) but into two uncoupled first order differential equations, which both are quite familiar. The second equation for the phase φ can readily be solved, $\varphi(t) = \omega t$, the phase is linearly increasing with time, and, as φ is a cyclic quantity, has to be taken modulo 2π. The first equation is the well-known cubic equation (7) this time simply written in the variable r instead of x. As we have seen earlier, this equation has a single stable fixed point $r = 0$ for $\varepsilon < 0$ and undergoes a pitchfork bifurcation at $\varepsilon = 0$, which turns the fixed point $r = 0$ unstable and gives rise to two new stable fixed points at $r = \pm\sqrt{\varepsilon}$. Interpreting r as the radius of the limit cycle, which has to be greater than zero, we find that a stable limit cycle arises from a fixed point, when ε exceeds its critical value $\varepsilon = 0$.

To characterize the behavior that a stable fixed point switches stability with a limit cycle in a more general way, we have a look at the linear part of (35) in its cartesian form

$$\dot{\xi} = \mu\xi = (\varepsilon + i\omega)(x + iy) \quad \rightarrow \quad \begin{pmatrix} \dot{x} \\ \dot{y} \end{pmatrix} = \begin{pmatrix} \varepsilon & -\omega \\ \omega & \varepsilon \end{pmatrix} \begin{pmatrix} x \\ y \end{pmatrix} \quad (38)$$

The eigenvalues λ for the matrix in (38) are found from the characteristic polynomial

$$\begin{vmatrix} \varepsilon - \lambda & -\omega \\ \omega & \varepsilon - \lambda \end{vmatrix} = \lambda^2 - 2\varepsilon\lambda + \varepsilon^2 + \omega^2$$

$$\rightarrow \quad \lambda_{1,2} = \varepsilon \pm \tfrac{1}{2}\sqrt{4\varepsilon^2 - 4\varepsilon^2 - 4\omega^2} = \varepsilon \pm i\omega \quad (39)$$

A plot of $\Im(\lambda)$ versus $\Re(\lambda)$ is shown in fig. 22 for the system we discussed here on the left, and for a more general case on the right. Such a qualitative change in a dynamical system where a pair of complex conjugate eigenvalues crosses the vertical axis we call a *Hopf bifurcation*, which is the most important bifurcation type for a system that switches from a stationary state at a fixed point to oscillation behavior on a limit cycle.

Dynamical Systems in One and Two Dimensions

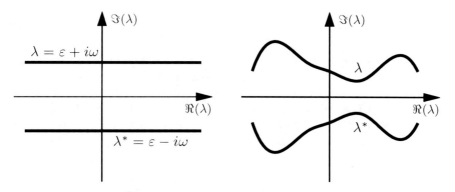

Fig. 22 A Hopf bifurcation occurs in a system when a pair of complex conjugate eigenvalues crosses the imaginary axis. For (39) the imaginary part of ε is a constant ω (left). A more general example is shown on the right.

2.5 Potential Functions in Two-Dimensional Systems

A two-dimensional system of first order differential equations of the form

$$\dot{x} = f(x,y) \qquad \dot{y} = g(x,y) \tag{40}$$

has a potential and is called a *gradient system* if there exists a scalar function of two variables $V(x,y)$ such that

$$\begin{pmatrix} \dot{x} \\ \dot{y} \end{pmatrix} = \begin{pmatrix} f(x,y) \\ g(x,y) \end{pmatrix} = - \begin{pmatrix} \dfrac{\partial V(x,y)}{\partial x} \\ \dfrac{\partial V(x,y)}{\partial y} \end{pmatrix} \tag{41}$$

is fulfilled. As in the one-dimensional case the potential function $V(x,y)$ is monotonically decreasing as time evolves, in fact, the dynamics follows the negative gradient and therefore the direction of steepest decent along the two-dimensional surface This implies that a gradient system cannot have any closed orbits or limit cycles.

An almost trivial example for a two-dimensional system that has a potential is given by

$$\dot{x} = -\frac{\partial V}{\partial x} = -x \qquad \dot{y} = -\frac{\partial V}{\partial x} = y \tag{42}$$

Technically, (42) is not even two-dimensional but two one-dimensional systems that are uncoupled. The eigenvalues and eigenvectors can easily be guessed as $\lambda_1 = -1$, $\lambda_2 = 1$ and $\mathbf{v}^{(1)} = (1,0)$, $\mathbf{v}^{(2)} = (0,1)$ defining the x-axis as a stable and the y-axis as an unstable direction. Applying the classification scheme, with $t_r = 0$ and $d_{et} = -1$ the origin is identified as a saddle. It is also easy to guess the potential function $V(x,y)$ for (42) and verify the guess by taking the derivatives with respect to x and y

$$V(x,y) = \frac{1}{2}x^2 - \frac{1}{2}y^2 \quad \rightarrow \quad \frac{\partial V}{\partial x} = x = -\dot{x} \quad \frac{\partial V}{\partial y} = -y = -\dot{y} \quad (43)$$

A plot of this function in shown in fig. 23 (left). White lines indicate equipotential locations and a set of trajectories is plotted in black. The trajectories are following the negative gradient of the potential and therefore intersect the equipotential lines at a right angle. From the shape of the potential function on the left it is most evident why fixed points in two dimensions with a stable and an unstable direction are called saddles.

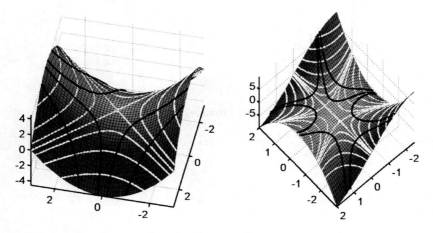

Fig. 23 Potential functions for a saddle (42) (left) and for the example given by (46) (right). Equipotential lines are plotted in white and a set of trajectories in black. As the trajectories follow the negative gradient of the potential they intersect the lines of equipotential at a right angle.

It is easy to figure out whether a specific two-dimensional system is a gradient system and can be derived from a scalar potential function. A theorem states that a potential exists if and only if the relation

$$\frac{\partial f(x,y)}{\partial y} = \frac{\partial g(x,y)}{\partial x} \quad (44)$$

is fulfilled. We can easily verify that (42) fulfills this condition

$$\frac{\partial f(x,y)}{\partial y} = -\frac{\partial x}{\partial y} = \frac{\partial g(x,y)}{\partial x} = \frac{\partial y}{\partial x} = 0 \quad (45)$$

However, in contrast to one-dimensional systems, which all have a potential, two-dimensional gradient systems are more the exception than the rule.

As a second and less trivial example we discuss the system

$$\dot{x} = y + 2xy \qquad \dot{y} = x + x^2 - y^2 \quad (46)$$

First we check whether (44) is fulfilled and (46) can indeed be derived from a potential

$$\frac{\partial f(x,y)}{\partial y} = \frac{\partial (y+2xy)}{\partial y} = \frac{\partial g(x,y)}{\partial x} = \frac{\partial (x+x^2-y^2)}{\partial x} = 1+2x \qquad (47)$$

In order to find the explicit form of the potential function we first integrate $f(x,y)$ with respect to x, and $g(x,y)$ with respect to y

$$\begin{aligned}\dot{x} = f(x,y) = -\frac{\partial V}{\partial x} &\rightarrow V(x,y) = -xy - x^2y + c_x(y)\\ \dot{y} = g(x,y) = -\frac{\partial V}{\partial y} &\rightarrow V(x,y) = -xy - x^2y + \tfrac{1}{3}y^3 + c_y(x)\end{aligned} \qquad (48)$$

As indicated the integration "constant" c_x for the x integration is still dependent on the the variable y and vice versa for c_y. These constants have to be chosen such that the potential $V(x,y)$ is the same for both cases, which is evidently fulfilled by choosing $c_x(y) = \tfrac{1}{3}y^3$ and $c_y(x) = 0$. A plot of $V(x,y)$ is shown in fig. 23 (right). Equipotential lines are shown in white and some trajectories in black. Again the trajectories follow the gradient of the potential and intersect the contour lines at a right angle.

2.6 Oscillators

Harmonic Oscillator

The by far best known two-dimensional dynamical system is the harmonic oscillator given by

$$\ddot{x} + 2\gamma\dot{x} + \omega^2 x = 0 \quad \text{or} \quad \begin{cases} \dot{x} = y \\ \dot{y} = -2\gamma y - \omega^2 x \end{cases} \qquad (49)$$

Here ω is the angular velocity, sometimes referred to in a sloppy way as frequency, γ is the damping constant and the factor 2 allows for avoiding fractions in some formulas later on. If the damping constant vanishes, the trace of the linear matrix is zero and its determinant ω^2, which classifies the fixed point at the origin as a center. The generals solution of (49) in this case is given by a superposition of a cosine and sine function

$$x(t) = a\cos\omega t + b\sin\omega t \qquad (50)$$

where the open parameters a and b have to be determined from initial conditions, displacement and velocity at $t = 0$ for instance.

If the damping constant is finite, the trance longer vanishes and the phase space portrait is a stable or unstable spiral depending on the sign of γ. For $\gamma > 0$ the time series is a damped oscillation (unless the damping gets really big, a case we leave as an exercise for the reader) and for $\gamma < 0$ the amplitude increases exponentially,

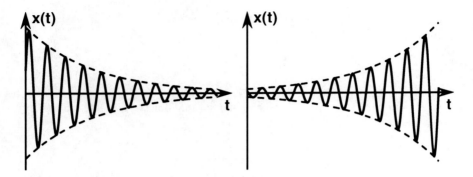

Fig. 24 Examples for "damped" harmonic oscillations for the case of positive damping $\gamma > 0$ (left) and negative damping $\gamma < 0$ (right).

both cases are shown in fig. 24. As it turns out, the damping not only has an effect on the amplitude but also on the frequency and the general solution of (49) reads

$$x(t) = e^{-\gamma t}\{a\cos\Omega t + b\sin\Omega t\} \quad \text{with} \quad \Omega = \sqrt{\gamma^2 - \omega^2} \tag{51}$$

Nonlinear Oscillators

As we have seen above harmonic (linear) oscillators do not have limit cycles, i.e. *isolated* closed orbits in phase space. For the linear center, there is always another orbit infinitely close by, so if a dynamics is perturbed it simply stays on the new trajectory and does not return to the orbit it originated from. This situation changes drastically as soon as we introduce nonlinear terms into the oscillator equation

$$\ddot{x} + \gamma\dot{x} + \omega^2 x + N(x,\dot{x}) = 0 \tag{52}$$

For the nonlinearites $N(x,\dot{x})$ there are infinitely many possibilities, even if we restrict ourselves to polynomials in x and $\dot{x}$. However, depending on the application there are certain terms that are more important than others, and certain properties of the system we are trying to model may give us hints, which nonlinearities to use or to exclude.

As an example we are looking for a nonlinear oscillator to describe the movements of a human limb like a finger, hand, arm or leg. Such movements are indeed limit cycles in phase space and if their amplitude is perturbed they return to the formerly stable orbit. For simplicity we assume that the nonlinearity is a polynomial in x and $\dot{x}$ up to third order, which means we can pick from the terms

$$\begin{aligned}\text{quadratic:} &\quad x^2, x\dot{x}, \dot{x}^2 \\ \text{cubic:} &\quad x^3, x^2\dot{x}, x\dot{x}^2, \dot{x}^3\end{aligned} \tag{53}$$

For human limb movements, the flexion phase is in good approximation a mirror image of the extension phase. In the phase space portrait this is reflected by a point

Dynamical Systems in One and Two Dimensions

symmetry with respect to the origin or an invariance of the system under the transformation $x \to -x$ and $\dot{x} \to -\dot{x}$. In order to see the consequences of such an invariance we probe the system

$$\ddot{x} + \gamma\dot{x} + \omega^2 x + ax^2 + bx\dot{x} + cx^3 + dx^2\dot{x} = 0 \tag{54}$$

In (54) we substitute x by $-x$ and $\dot{x}$ by $-\dot{x}$ and obtain

$$-\ddot{x} - \gamma\dot{x} - \omega^2 x + ax^2 + bx\dot{x} - cx^3 - dx^2\dot{x} = 0 \tag{55}$$

Now we multiply (55) by -1

$$\ddot{x} + \gamma\dot{x} + \omega^2 x - ax^2 - bx\dot{x} + cx^3 + dx^2\dot{x} = 0 \tag{56}$$

Comparing (56) with (54) shows that the two equations are identical if and only if the coefficients a and b are zero. In fact, evidently any quadratic term cannot appear in an equation for a system intended to serve as a model for human limb movements as it breaks the required symmetry. From the cubic terms the two most important ones are those that have a main influence on the amplitude as we shall discuss in more details below. Namely, these nonlinearities are the so-called *van-der-Pol* term $x^2\dot{x}$ and the *Rayleigh* term $\dot{x}^3$.

Van-der-Pol Oscillator: $N(x, \dot{x}) = x^2\dot{x}$

The van-der-Pol oscillator is given by

$$\ddot{x} + \gamma\dot{x} + \omega^2 x + \varepsilon x^2 \dot{x} = 0 \tag{57}$$

which we can rewrite in the form

$$\ddot{x} + \underbrace{(\gamma + \varepsilon x^2)}_{\tilde{\gamma}}\dot{x} + \omega^2 x = 0 \tag{58}$$

Equation (58) shows that for the van-der-Pol oscillator the damping "constant" $\tilde{\gamma}$ becomes time dependent via the amplitude x^2. Moreover, writing the van-der-Pol oscillator in the form (58) allows for an easy determination of the parameter values for γ and ε that can lead to sustained oscillations. We distinguish four cases:

$\gamma > 0, \varepsilon > 0$: The effective damping $\tilde{\gamma}$ is always positive. The trajectories are evolving towards the origin which is a stable fixed point;

$\gamma < 0, \varepsilon < 0$: The effective damping $\tilde{\gamma}$ is always negative. The system is unstable and the trajectories are evolving towards infinity;

$\gamma > 0, \varepsilon < 0$: For small values of the amplitude x^2 the effective damping $\tilde{\gamma}$ is positive leading to even smaller amplitudes. For large values of x^2 the effective damping $\tilde{\gamma}$ is negative leading a further increase in amplitude. The system evolves either towards the fixed point or towards infinity depending on the initial conditions;

$\gamma < 0$, $\varepsilon > 0$: For small values of the amplitude x^2 the effective damping $\tilde{\gamma}$ is negative leading to an increase in amplitude. For large values of x^2 the effective damping $\tilde{\gamma}$ is positive and decreases the amplitude. The system evolves towards a stable limit cycle. Here we see a familiar scenario: without the nonlinearity the system is unstable ($\gamma < 0$) and moves away from the fixed point at the origin. As the amplitude increases the nonlinear damping ($\varepsilon > 0$) becomes an important player and leads to saturation at a finite value.

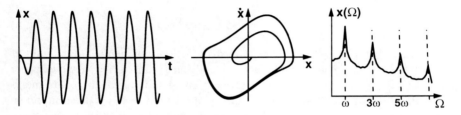

Fig. 25 The van-der-Pol oscillator: time series (left), phase space trajectory (middle) and power spectrum (right).

The main features for the van-der-Pol oscillator are shown in fig. 25 with the time series (left), the phase space portrait (middle) and the power spectrum (right). The time series is not a sine function but has a fast rising increasing flank and a more shallow slope on the decreasing side. Such time series are called *relaxation oscillations*. The trajectory in phase space is closer to a rectangle than a circle and the power spectrum shows pronounced peaks at the fundamental frequency ω and its odd higher harmonics ($3\omega, 5\omega \ldots$).

Rayleigh Oscillator: $\mathbf{N(x,\dot{x}) = \dot{x}^3}$

The Rayleigh oscillator is given by

$$\ddot{x} + \gamma \dot{x} + \omega^2 x + \delta \dot{x}^3 = 0 \tag{59}$$

which we can rewrite as before

$$\ddot{x} + \underbrace{(\gamma + \delta \dot{x}^2)}_{\tilde{\gamma}} \dot{x} + \omega^2 x = 0 \tag{60}$$

In contrast to the van-der-Pol case the damping "constant" for the Rayleigh oscillator depends on the square of the velocity $\dot{x}^2$. Arguments similar to those used above lead to the conclusion that the Rayleigh oscillator shows sustained oscillations in the parameter range $\gamma < 0$ and $\delta > 0$.

As shown in fig. 26 the time series and trajectories of the Rayleigh oscillator also show relaxation behavior, but in this case with a slow rise and fast drop. As for

Dynamical Systems in One and Two Dimensions

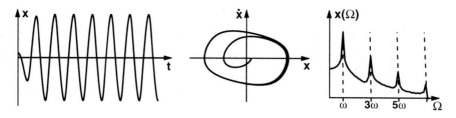

Fig. 26 The Rayleigh oscillator: time series (left), phase space trajectory (middle) and power spectrum (right).

the van-der-Pol oscillator, the phase space portrait is almost rectangular but the long and short axes are switched. Again the power spectrum has peaks at the fundamental frequency and the odd higher harmonics.

Taken by themselves neither the van-der-Pol nor Rayleigh oscillators are good models for human limb movement for at least two reasons even though they fulfill one requirement for a model: they have stable limit cycles. However, first, human limb movements are almost sinusoidal and their trajectories have a circular or elliptic shape. Second, it has been found in experiments with human subjects performing rhythmic limb movements that when the movement rate is increased, the amplitude of the movement decreases linearly with frequency. It can be shown that for the van-der-Pol oscillator the amplitude is independent of frequency and for the Rayleigh it decreases proportional to ω^{-2}, both in disagreement with the experimental findings.

Hybrid Oscillator: $N(x, \dot{x}) = \{x^2 \dot{x}, \dot{x}^3\}$

The hybrid oscillator has two nonlinearities, a van-der-Pol and a Rayleigh term and is given by

$$\ddot{x} + \gamma \dot{x} + \omega^2 x + \varepsilon x^2 \dot{x} + \delta \dot{x}^3 = 0 \tag{61}$$

which we can rewrite again

$$\ddot{x} + \underbrace{(\gamma + \varepsilon x^2 + \delta \dot{x}^2)}_{\tilde{\gamma}} \dot{x} + \omega^2 x = 0 \tag{62}$$

The parameter range of interest is $\gamma < 0$ and $\varepsilon \approx \delta > 0$. As seen above, the relaxation phase occurs on opposite flanks for the van-der-Pol and Rayleigh oscillator. In combining both we find a system that not only has a stable limit cycle but also the other properties required for a model of human limb movement.

As shown in fig. 27 the time series for the hybrid oscillator is almost sinusoidal and the trajectory is elliptical. The power spectrum has a single peak at the fundamental frequency. Moreover, the relation between the amplitude and frequency is a linear decrease in amplitude when the rate is increased as shown schematically in fig. 28. Taken together, the hybrid oscillator is a good approximation for the trajectories of human limb movements.

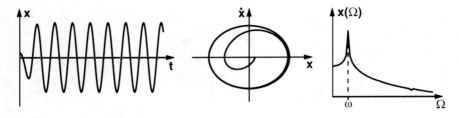

Fig. 27 The hybrid oscillator: time series (left), phase space trajectory (middle) and power spectrum (right).

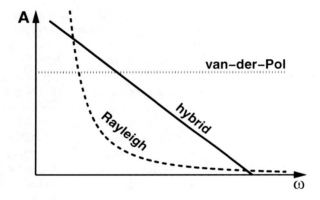

Fig. 28 Amplitude-frequency relation for the van-der-Pol (dotted), Rayleigh ($\sim \omega^{-2}$, dashed) and hybrid ($\sim -\omega$, solid) oscillator.

Beside the dynamical properties of the different oscillators, the important issue here, which we want to emphasize on, is the modeling strategy we have applied. Starting from a variety of quadratic and cubic nonlinearities in x and $\dot{x}$ we first used the symmetry between the flexion and extension phase of the movement to rule out any quadratic terms. Then we studied the influence of the van-der-Pol and Rayleigh terms on the time series, phase portraits and spectra. In combining these nonlinearities to the hybrid oscillator we found a dynamical system that is in agreement with the experimental findings, namely

- the trajectory in phase space is a stable limit cycle. If this trajectory is perturbed the system returns to its original orbit;
- the time series of the movement is sinusoidal and the phase portrait is elliptical;
- the amplitude of the oscillation decreases linearly with the movement frequency.

For the sake of completeness we briefly mention the influence of the two remaining cubic nonlinearities on the dynamics of the oscillator. The van-der-Pol and Rayleigh term have a structure of velocity times the square of location or velocity, respectively, which we have written as a new time dependent damping term. Similarly, the remaining terms $x\dot{x}^2$ and x^3 (the latter called a Duffing term) are of the form location times the square of velocity or location. These nonlinearities can be written as a time dependent frequency, leading to an oscillator equation with all cubic nonlinear terms

$$\ddot{x} + \underbrace{(\gamma + \varepsilon x^2 + \delta \dot{x}^2)}_{\tilde{\gamma}\text{ damping}} \dot{x} + \underbrace{(\omega^2 + \alpha \dot{x}^2 + \beta x^2)}_{\tilde{\omega}^2\text{ frequency}} x = 0 \qquad (63)$$

Further Readings

Strogatz, S.H.: Nonlinear Dynamics and Chaos. Perseus Books Publishing, Cambridge (2000)
Haken, H.: Introduction and Advanced Topics. Springer, Berlin (2004)

Benefits and Pitfalls in Analyzing Noise in Dynamical Systems – On Stochastic Differential Equations and System Identification

Andreas Daffertshofer

1 Introduction

The search for a mathematical framework for describing motor behavior has a long but checkered history. Most studies have focused on recurrent, deterministic features of behavior. The use of dynamical systems to account for the qualitative features of end-effector trajectories of limb oscillations gained momentum in the last twenty-five years or so. There, salient characteristics of human movement served as guidelines for model developments. For instance, trajectories of limb cycling describing a bounded area in the position-velocity or phase plane may be interpreted as indicative of a limit cycle attractor, at least when modeling efforts are restricted to identifying deterministic forms, thereby disregarding variability. By using averaging methods, which are typically applied for first-order analyses of nonlinear oscillators, e.g., harmonic balance, Kay et al (1987, 1991) derived second-order nonlinear differential equations that mimic experimentally observed amplitude-frequency relations and phase response characteristics of rhythmic finger and wrist movements. The self-sustaining oscillators include weak dissipative nonlinearities that stabilize the underlying limit cycle, cause a drop of amplitude and an increase in peak velocity with increasing movement tempo or frequency.

Random noise is omnipresent in motor behavior but noise in limb oscillations, although explicitly acknowledged by at least some authors (e.g., Eisenhammer et al, 1991; Kay, 1988), is viewed as mere artifact obscuring the deterministic dynamics. Noise should thus be eliminated by means of filtering or averaging. In consequence, deterministic and stochastic features of human movement are rarely assessed in conjunction. Fortunately, more recently studies are on the rise, which do appreciate the novel dynamic characteristics caused by noise in the dynamics. In these more recent studies, stochasticity is considered a hallmark property of human

---
Andreas Daffertshofer
Research Institute MOVE, VU University Amsterdam
van der Boechorststraat 9, 1081 BT, Amsterdam, The Netherlands
e-mail: `marlow@fbw.vu.nl`

movement that needs to be addressed because it possesses functional qualities (e.g., Harris and Wolpert, 1998; Körding and Wolpert, 2004; Riley and Turvey, 2002; Schöner et al, 1986; Schöner, 2002). For instance, variability in endpoint trajectory has been associated with task difficulty (Todorov and Jordan, 2002): variability is reduced in more difficult tasks to comply with accuracy constraints, whereas in easier tasks the variability can increase in order to enhance system flexibility. Another example of the usefulness of motor variability can be found in studies of interlimb coordination conducted from a dynamical systems perspective. In this context, variability has been incorporated as random fluctuations to account for phenomena like critical fluctuations and critical slowing down in the immediate vicinity of so-called phase transitions, that is, situations in which a system switches between stable states or attractors, e.g., changes from antiphase to in-phase coordination (Haken et al, 1985; Kelso, 1984; Post et al, 2000; Schöner et al, 1986, and the *Chapter* by Calvin and Jirsa in the current volume). Put differently, random fluctuations seem to compete with stability and thus contribute to the flexibility of a system. Strong fluctuations reflect less stable states between which the system may readily switch, whereas weak fluctuations indicate more stable states that can be steadily maintained. Formally, variability may be accounted for by incorporating either additive or multiplicative random fluctuations yielding stochastic differential equations. Schöner et al (1986) formulated stochastic models similar to the above-mentioned deterministic limit cycle models under the impact of stochastic forces. Schöner (1990) also suggested similar accounts of discrete movement and postural sway, albeit without providing any empirical support.

Here, the mathematical framework of stochastic differential equations will be briefly discussed. Several examples ranging from classic physical problems to human movement will be highlighted. Reducing the description of the dynamics to that of a single (scalar) variable, a periodic forcing will be included to illustrate the phenomenon of stochastic resonance. Finally, a recently established analysis method of system identification will be described that allows for an unbiased specification of deterministic and stochastic system components. Its expediency is illustrated in analyzing kinematic data of rhythmic tapping. The ultimate goal of this analysis is to find mathematical descriptions that exhibit the main dynamic features of the system under study.

2 Probability

Stochastic behavior can be considered as the time evolution of a process under the impact of random fluctuations. To express this behavior, a mathematical description that allows for predicting behavior given a current state is required. By definition, however, the future states of a stochastic process cannot be fully determined but only estimated to some extent. That is, any prediction is conditional upon the probability that a certain future state will occur or to find the system's state at a certain instant in time in a specific area in state space. A collection of such probabilities, which indicate the likelihood of the occurrence of specific states is called a probability distribution function.

Throughout this *Chapter* the probability distribution corresponding to a continuous stochastic dynamics will be defined via a (virtual) ensemble of identical systems meaning that, e.g., the mean, variance, or higher order cumulants are computed over this ensemble of systems rather than over the evolution of a single realization. Put differently, the probability distribution is given via a distribution of a collection of these identical systems that intermingle as a result of their spontaneous evolution influenced by random fluctuations.

2.1 Mean and Expectation Values

If an ensemble of stochastic variables is known, one can index any specific realization of the ensemble by ξ_k. This yields the mean value over all realizations as

$$\overline{\xi} = \frac{1}{N} \sum_{k=1}^{N} \xi_k. \tag{1}$$

N denotes the number of realizations. Equivalently, the mean of an arbitrary function f of ξ reads

$$\overline{f(\xi)} = \frac{1}{N} \sum_{k=1}^{N} f(\xi_k). \tag{2}$$

For the sake of convenience, we consider the case in which the ensemble is infinitely large, i.e. we look at the limit $N \to \infty$. As will be shown below, in this limit the mean (2) agrees with the expectation value of the stochastic variable ξ, which can hence be given as

$$\mathbb{E}[f(\xi)] = \lim_{N \to \infty} \frac{1}{N} \sum_{k=1}^{N} f(\xi_k). \tag{3}$$

2.2 Probability Distribution and Density

The probability distribution function Pr_ξ of a system that relates to the stochastic variable ξ can be readily defined using the so-called Heaviside function, or unit step function, which is given by $\Theta(x) = 0$ for all $x < 0$ and 1 otherwise. With this function the number M of realizations ξ_k that are smaller or equal to a fixed value x can be written as $M = \sum_k \Theta(x - \xi_k)$. Accordingly, the probability to find a realization ξ being smaller or equal than x in an ensemble of N realizations follows as

$$Pr_\xi(\xi \leq x) = \frac{M}{N} = \frac{1}{N} \sum_{k=1}^{N} \Theta(x - \xi_k),$$

or, if looking again at the limit $N \to \infty$,

$$Pr_\xi\left(\xi \leq x\right) = \lim_{N\to\infty} \frac{1}{N}\sum_{k=1}^{N}\Theta\left(x-\xi_k\right) = \mathbb{E}\left[\Theta\left(x-\xi_k\right)\right].$$

In consequence, the probability for a realization to fall in the interval $x < \xi \leq x+\Delta x$ is given as difference

$$Pr_\xi\left(x < \xi \leq x+\Delta x\right) = \mathbb{E}\left[\Theta\left(x+\Delta x-\xi\right) - \Theta\left(x-\xi\right)\right], \tag{4}$$

which, in turn, defines the probability distribution function Pr_ξ. Converting this difference to a differential, i.e. considering $\Delta x \to dx$ to be infinitesimal, yields the probability density function p_ξ

$$p_\xi\left(x\right) = \frac{\partial}{\partial x}Pr_\xi\left(\xi \leq x\right) = \frac{\partial}{\partial x}\mathbb{E}\left[\Theta\left(x-\xi\right)\right] = \mathbb{E}\left[\frac{\partial \Theta\left(x-\xi\right)}{\partial x}\right] = \mathbb{E}\left[\delta\left(x-\xi\right)\right] \tag{5}$$

that provides the probability to find the stochastic variable ξ in the infinitesimal interval of state values $[x, x+dx]$. Here, Pr_ξ denotes the probability distribution as is common in the mathematics literature. In physics, however, p_ξ is often also called distribution sometimes causing confusion. Therefore, we here always refer to p_ξ as probability density function.

In Eq. (5) δ refers to Dirac's δ-distribution, which has the interesting properties: $\delta(x) = 0$ for all $x \neq 0$ and $\int_{-\varepsilon}^{\varepsilon}\delta(x)dx = 1$ for all $\varepsilon > 0$, and

$$\int f\left(x\right)\delta\left(x-y\right)dx = f\left(y\right).$$

Note that these properties reveal the consistency of Eq. (3) when starting off the conventional definition of the expectation value by means of

$$\mathbb{E}\left[f\left(\xi\right)\right] = \int f\left(x\right)p_\xi\left(x\right)dx = \int f\left(x\right)\mathbb{E}\left[\delta\left(x-\xi\right)\right]dx = \int f\left(x\right)\left[\lim_{N\to\infty}\frac{1}{N}\sum_{k=1}^{N}\delta\left(x-\xi_k\right)\right]dx$$
$$= \lim_{N\to\infty}\frac{1}{N}\sum_{k=1}^{N}\int f\left(x\right)\delta\left(x-\xi_k\right)dx = \lim_{N\to\infty}\frac{1}{N}\sum_{k=1}^{N}f\left(\xi_k\right).$$

2.2.1 Moments and Cumulants

A probability distribution and its density function can be determined by their corresponding statistical characteristics like its moments or its cumulants. The cumulants relate to the so-called characteristic function of the probability density, i.e. its Fourier transform, that provides an alternative description of the stochastic process of interest (see Risken, 1989, Chap 2.2 for more details). In the one-dimensional case the moments are directly and the cumulants recursively defined via

$$m_k = \int x^k p_\xi\left(x\right)dx = \mathbb{E}\left[\xi^k\right] \quad \text{and} \quad c_1 = m_1 \quad \text{and for} \quad k > 1$$
$$c_k = \mathbb{E}\left[\left(\xi - m_1\right)^k\right] - \sum_{l=1}^{k-1}\binom{k-1}{l-1}c_l\mathbb{E}\left[\left(\xi - m_1\right)^{k-l}\right]$$

respectively. The first cumulant is hence equal to the mean, the second is the variance, i.e. the second centered moment, the third is the third centered moment that relates to the skewness, and so on[1]. For the sake of completeness we list the characteristic function that reads

$$\int e^{ixy} p_\xi(x)\, dx = \mathbb{E}\left[e^{i\xi y}\right] = 1 + \sum_{k=1}^{\infty} \frac{(iy)^k}{k!} m_k = \exp\left\{\sum_{k=1}^{\infty} \frac{(iy)^k}{k!} c_k\right\}.$$

3 Dynamics and Noise

Before discussing the dynamics of the probability density function the (more intuitive) notion of dynamics under influence of random fluctuations will be defined. Let us consider an arbitrary, real-valued, random variable describing the system under study and denote it by ξ. This variable can be the n-dimensional representation of the system. For the sake of legibility, however, here we restrict the notation to scalar values (more general descriptions can be found in, e.g., Gardiner, 2004; Risken, 1989).

3.1 Stochastic Dynamics

In the case of a dynamical system, the stochastic variable is identified as a function of time in terms of $\xi = \xi(t)$. We discuss dynamical systems that can be cast in the form of a so-called generalized Langevin equation meaning that the time evolution of ξ can be formally written as[2]

$$\dot{\xi}(t) = f(\xi(t), t) + g(\xi(t), t) \Gamma(t). \tag{6}$$

The function f combines all deterministic forces acting on ξ, and the function g represents possible state dependent forms that may modulate random fluctuations $\Gamma(t)$ before they are incorporated into the dynamics. The dot-notation refers to the derivative with respect to time t. The functions f and g may also explicity depend on time but unless stated otherwise this case will here not further pursued.

In dynamical systems, noise can be incorporated in (at least) two different ways. Either the noise is just added to the state variable but does not influence the dynamics – that case is commonly viewed as so-called measurement noise and will not be considered here; see (Siefert et al, 2003) for an in-depth discussion. More interesting is the case in which the noise is an intrinsic part of the dynamics, i.e. it is included in the system's evolution as, for instance, shown in (6). This intrinsic randomness either multiplies or just adds into the dynamics, which means in (6) that the function g explicitly depends on ξ or not, respectively. Moreover, one discriminates

---

[1] Note that a normal or Gaussian distribution is completely specified by its mean and variance.
[2] ξ is a stochastic variable and not necessarily differentiable. Here, the temporal derivative is just a symbolic notion; see Gardiner (2004) for more details.

between uncorrelated (or white) and correlated (or colored) noise. The latter, however, simply entails the presence of a supplementary deterministic component that generates some correlations in an otherwise random process. For example, exponentially correlated noise can be expressed as linearly damped dynamics systems with additive uncorrelated noise; see below in *Paragraph 3.1.1*. Put differently, correlated or colored noise can be seen as the presence of an auxiliary (or hidden) variable that follows a certain deterministic dynamics under impact of uncorrelated, white noise. In the present *Chapter*, however, all the deterministic forces are combined in the function $f(\xi)$. By this we can restrict ourselves to the discussion of uncorrelated noise. Without loss of generality, we also assume that the noise values are normally distributed and have vanishing mean. In combination we therefore always consider so-called mean-centered Gaussian white noise, for which one writes

$$\mathbb{E}\left[\Gamma(t)\right] = 0 \quad \text{and} \quad \mathbb{E}\left[\Gamma(t)\Gamma(t')\right] = Q\delta(t - t'). \tag{7}$$

Before investigating the dynamics of the probability density corresponding to Eq. (6), the averaged solutions of the two seminal examples will be discussed: Brownian motion and the so-called Wiener process. These examples already provide a feel for the underlying mathematical forms and the expected outcome for the subsequently deduced, more general description. As indicated above, we consider (infinitely) many realizations of a specific system and compute the expectation value over all these realizations. To investigate temporal characteristics of the system, we further compute its temporal correlations, e.g., the auto-correlation function or the closely related least-squared displacement.

3.1.1 Brownian Motion

The most well-known stochastic dynamical system is the one describing Brownian motion, i.e. the time evolution of non-interacting particles, each with mass m, damped by a linear force with strength b, and under influence of thermal noise $\tilde{\Gamma}(t)$. Using ψ as state variable and $\tilde{\Gamma}(t)$ for the noise, the dynamics may be written as

$$m\ddot{\psi}(t) + b\dot{\psi}(t) = \tilde{\Gamma}(t). \tag{8}$$

That equation can be transformed into the form of (6) when substituting $\xi = \dot{\psi}$, $c = b/m$, and $\sqrt{2Q}\Gamma(t) = (1/m)\tilde{\Gamma}(t)$. This leads to

$$\dot{\xi}(t) = -c\xi(t) + \sqrt{2Q}\Gamma(t). \tag{9}$$

For the general functions in the dynamics (6) this implies f to be linear, i.e. $f(\xi) = -c\xi$, and g to be constant and equal to $\sqrt{2Q}$, i.e. the noise is additive[3]. The square root notation in the dynamics is a convenient way to introduce the correlation strength $2Q$ of the included noise, which here equals its variance. The solution of Eq. (9) formally reads

---

[3] This means additive in the dynamics, not a mere addition to the state variable.

Analyzing Noise in Dynamical Systems

$$\xi(t) = \xi_0 e^{-ct} + \int_{t_0}^{t} \Gamma(\tau) e^{-c(t-\tau)} d\tau,$$

where ξ_0 is given by some initial condition. That said we now consider the aforementioned many realizations of such a system – think of the many non-interacting particles – and use definition (3), which provides the average evolution of an ensemble of Brownian particles as

$$\mathbb{E}[\xi(t)] = \xi_0 e^{-ct} + \underbrace{\mathbb{E}\left[\int_{0}^{t} \Gamma(\tau) e^{-c(t-\tau)} d\tau\right]}_{=\int_{0}^{t} \mathbb{E}[\Gamma(\tau)] e^{-c(t-\tau)} d\tau = 0} = \xi_0 e^{-ct}. \qquad (10)$$

In words, the expectation value $\mathbb{E}[\xi]$ evolves like an individual member of the ensemble in the absence of noise ($Q = 0$), i.e. in the purely deterministic version of (9). More interestingly, however, the system's auto-correlation is found as

$$\mathbb{E}[\xi(t + \Delta t) \xi(t)] = \frac{Q}{2c} e^{-c\Delta t};$$

note that for the sake of simplicity we put the initial value ξ_0 to zero. That is, Brownian motion has an exponentially decaying expectation value and an exponentially decaying auto-correlation; the latter can be used to, e.g., generate exponentially correlated noise (see above for the discussion on white viz. colored noise). Of course, these exponential decays merely reflect that the dynamics under study is linear (see also *Appendices A & B*).

3.1.2 Wiener Process

An even 'simpler' example than Brownian motion is that of a dynamics that does not contain any deterministic force. Since we consider uncorrelated Gaussian noise, such a dynamics may be written as

$$\dot{\xi} = \Gamma(t) \quad \text{or} \quad \xi(t) = \int_{0}^{t} \Gamma(\tau) d\tau, \qquad (11)$$

which describes a non-stationary stochastic process, i.e. the integral over white noise, referred to as Wiener process. The non-stationarity can be immediately realized by looking again at the temporal correlation structure of the process. In view of the non-stationarity it turns out to be somewhat simpler to study the least-squared displacement than the auto-correlation function, as it measures the diffusiveness (or spread) of the increments of a stochastic system. This least-squared displacement explicitly reads

$$\mathbb{E}\left[\{\xi(t+\Delta t) - \xi(t)\}^2\right] = \mathbb{E}\left[\{\xi(t)\}^2\right] + \mathbb{E}\left[\{\xi(t+\Delta t)\}^2\right]$$
$$- 2\underbrace{\mathbb{E}[\xi\{t+\Delta t\}\xi(t)]}_{=\text{auto-correlation}} = Q\Delta t. \qquad (12)$$

In words, every ensemble of different realizations of a Wiener process spreads linearly in time. This linear growth of the least-squared displacement is the most simple form of diffusion, which can be addressed in very general terms using the evolution of the corresponding probability distribution.

3.2 Time-Dependent Probability Density Functions

As mentioned above, the probability density function can be characterized by various statistical quantities, like its moments or its cumulants. For instance, if the process is nothing but diffusive, then the first two cumulants – mean and variance – fully determine the dynamics of the probability density while higher order cumulants turn out to be irrelevant, i.e., there are no jumps in the evolution of the probability density (see, e.g., Honerkamp, 1998, chap 5.6). Human movement can frequently be characterized as (the result of) such a diffusion process indeed, because it can often be captured in the form of the common stochastic differential equations (6), that is, a dynamical system (or differential forms) comprised of distinct deterministic and stochastic components. The unique link between these deterministic and stochastic components and the first two cumulants of the corresponding probability density function is well documented (Gardiner, 2004; Risken, 1989; Stratonovich, 1963). In fact, this link has provided a theoretical framework for a rigorous understanding of the interactions between deterministic and random features in complex dynamical systems (Haken, 1974).

The time dependence of the stochastic variable ξ suggests that the corresponding probability density is, in general, also time-dependent that can be formalized in terms of $\mathbb{E}\left[\delta\left(x - \xi\left(t\right)\right)\right] = p_{\xi(t)}(x) = p(x,t)$. For the sake of legibility, we drop the sub-script $_{\xi(t)}$ and only note that the evolution can be expanded by means of

$$p(x, t + \Delta t) = p(x,t) - \mathbb{E}\left[\frac{\partial \delta\left(x - \xi\left(t\right)\right)}{\partial x} \Delta \xi\right] + \frac{1}{2} \mathbb{E}\left[\frac{\partial^2 \delta\left(x - \xi\left(t\right)\right)}{\partial x^2} (\Delta \xi)^2\right] - \cdots$$

After some lengthy mathematical derivations (Gardiner, 2004; Kramers, 1940; Moyal, 1949), this expansion can be transformed into a dynamics[4]

$$\dot{p}(x, t | x_0, t_0) = \sum_{k=1}^{\infty} \frac{1}{k!} \left(-\frac{\partial}{\partial x}\right)^k \left[D^{(k)}(x) \, p(x, t | x_0, t_0)\right] \quad (13)$$

which provides a full description of the evolution of probability density function $p(x,t)$. Here, we abbreviated the so-called Kramers-Moyal coefficients

$$D^{(k)}(x) = \lim_{\Delta t \to 0} \frac{1}{\Delta t} \mathbb{E}\left[\{\xi(t + \Delta t) - \xi(t)\}^k\right] \quad (14)$$

---

[4] We only consider Markov processes, i.e. processes that can be fully described by their two-time correlation function; see *Paragraph 4.2*.

that agree with the cumulants of the conditional (or transition) probability from a state at time t to a state at time $t + \Delta t$; see also in *Paragraphs 2.2.1 & 4.1*.

3.2.1 Diffusion Equation

Under many circumstances the expansion (13) converges very quickly. In fact, whenever the third-order cumulant $D^{(3)}$ vanishes, all higher-order terms immediately disappear (Pawula, 1967) so that the dynamics of $p(x,t)$ includes only the first two Kramers-Moyal coefficients. These initial coefficients are referred to the drift, $D^{(1)}$, and diffusion, $D^{(2)}$, coefficients. As the name of the latter already implies the dynamics reduces to a diffusion equation. Then, the dynamics is given by the so-called Fokker-Planck equation, which reads

$$\dot{p}(x,t|x_0,t_0) = -\frac{\partial}{\partial x}\left[D^{(1)}(x)p(x,t|x_0,t_0)\right] + \frac{1}{2}\frac{\partial^2}{\partial x^2}\left[D^{(2)}(x)p(x,t|x_0,t_0)\right]. \quad (15)$$

In this case, the coefficients $D^{(1)}$ and $D^{(2)}$ are of primary interest because, when substituting them into Eq. (6), the stochastic differential equation of the system under study can be rewritten as

$$\dot{\xi} = D^{(1)}(\xi) + \sqrt{2D^{(2)}(\xi)}\,\Gamma(t) \quad (16)$$

How can these coefficients be determined given a set of empirical data? Before sketching a method to determine drift and diffusion coefficients in *Paragraph 4* and, by this, estimating the dynamics (16), the following sections concisely illustrate the relationship between the Langevin equation and the Fokker-Planck description by discussing dynamical models that are frequently discussed in the context of coordination dynamics.

3.3 Example: The HKB-Model

Kelso (1984) reported a by now paradigmatic experiment on rhythmic finger movements, which demonstrated the occurrence of phase transitions in human interlimb coordination. When subjects start out to cycle their index fingers (or hands) rhythmically in antiphase (simultaneous activation of nonhomologous muscle groups) and gradually increase the cycling frequency as prescribed by a metronome, a spontaneous, involuntary switch to the in-phase pattern (simultaneous activation of homologous muscle groups) occurs at a distinct, critical movement frequency. Beyond the critical frequency only the in-phase coordination can be stably performed. The observed change in coordination can be considered an instance of qualitative changes in macroscopic systems (here fingers) described by the theory of pattern formation in open systems far away from thermal equilibrium. Haken et al (1985) presented a theoretical model for the phase transitions of interest, now widely known as the HKB-model. Using the notion in Eq. (6), Schöner et al (1986) sketched its stochastic extension, which generally allows for studying the stability characteristics of the phase difference between the oscillating end-effectors in terms of its time-varying statistical properties as outlined above. In a nutshell, the deterministic part of the evolution can be described in terms of a gradient dynamics that involves a potential $V = V(\phi)$. With that the 'full' stochastic HKB-dynamics reads[5]

$$\dot{\xi}_\phi = -\frac{dV}{d\xi_\phi} + \sqrt{2Q}\Gamma(t) \quad \text{with} \quad V(\phi) = -\cos\phi - \frac{\varepsilon}{4}\cos 2\phi. \tag{17}$$

The minima of the potential $V(\phi)$, represent stable attractor states, here, the in-phase and antiphase coordination modes of the relative phase between the end-effectors; see also Eq. (35) below. ε serves as so-called control parameter, which has to be capable of inducing changes in V in order to account for the transition from anti- to in-phase coordination patterns. The critical value equals $\varepsilon = 1$, where the stable solution for antiphase coordination is annihilated; see Fig. 1, left panel.

With the results from *Paragraph 3.2.1*, the dynamics of the corresponding probability density function readily obtains the form

$$\dot{p}(\phi,t) = \frac{\partial}{\partial \phi}\left[\left\{\sin\phi + \frac{\varepsilon}{2}\sin 2\phi\right\}p(\phi,t)\right] + Q\frac{\partial^2}{\partial \phi^2}p(\phi,t), \tag{18}$$

which can indeed be solved analytically. For the stationary solution one finds

$$p_{\text{st}}(\phi) \propto e^{-V(\phi)/Q}; \tag{19}$$

$V(\phi)$ is given in the right hand side of (17). A sketch of this stationary density as function of the state variable and of the control parameter ε is shown in Fig. 1 (right panel) that clearly resembles the roughly inverted potential function V (left panel).

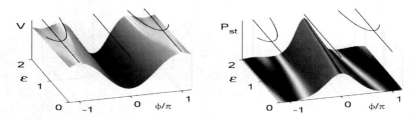

Fig. 1 HKB-model: left panel: potential $V = V(\phi)$ as a function of the state variable and the control parameter ε; the bifurcation occurs at $\varepsilon = 1$; right panel: corresponding stationary probability density also as function of the state variable and the control parameter; see text for further explanation.

---

[5] The mere addition of noise in the phase dynamics is, strictly speaking, an oversimplification when starting of an oscillator dynamics with additive noise, since then the noise in the phase dynamics is stronger for smaller amplitudes. The noise strength is inversely related with the amplitude of the oscillator; see Eq. (27).

In the HKB-potential the bifurcation parameter ε is considered to depend reciprocally on movement frequency. To formalize this relation, the HKB-model has been derived using a system of two coupled oscillators mimicking the moving limbs as limit cycle oscillators (Haken et al, 1985). Any smoothly evolving, deterministic oscillator can be formalized as second-order differential equation $\ddot{x} + \omega_0^2 x = n(x, \dot{x})$, or in its equivalent two-dimensional form of first-order equations listed in Eq. (20) below. The self-sustaining (autonomous) oscillators of limb movements contain next to a positive, linear component, also different negative, nonlinear damping terms (e.g., Rayleigh and van der Pol terms), which may generate specific dependency on the oscillator's amplitude and natural frequency ω_0. More important for the derivation of the HBK-model, however, is the choice of the coupling function between the two participating oscillators (see, e.g., Haken et al, 1985; Beek et al, 2002, for details). However, instead of pursuing this admittedly very important issue, we here focus on the dynamics of a single oscillator and investigate the effects of additive noise on its amplitude (and phase) dynamics.

3.4 *Spurious Drift in the Amplitude Dynamics of a Limit Cycle Oscillator*

Nonlinear oscillators have been frequently discussed in various scientific disciplines and their rigorous mathematical investigation goes back to the 19th century. Comparably new are studies of randomly forced limit cycles, though they also date back to the mid of the 20th century (Wax, 1954; Stratonovich, 1963; Has'minskiĭ, 1980). We concentrate on specific types of weakly nonlinear oscillators by writing

$$\frac{d}{dt}\begin{pmatrix} x \\ y \end{pmatrix} = \begin{pmatrix} 0 & 1 \\ -\omega_0^2 & 0 \end{pmatrix}\begin{pmatrix} x \\ y \end{pmatrix} + \begin{pmatrix} 0 \\ 1 \end{pmatrix} f(x, y) \qquad (20)$$

and including the aforementioned Rayleigh, Van der Pol, and Duffing nonlinearities by means of

$$f(x, y) = \omega_0 \left(\alpha - \frac{\beta}{3\omega_0^2} y^2 - \gamma x^2 \right) y - \frac{\omega_0^2 \eta}{3} x^3. \qquad (21)$$

Here $\beta \cdots$ refers to the Rayleigh, $\gamma \cdots$ to the van der Pol, and $\eta \cdots$ to the Duffing component[6]. The stochastic extension is, as usual in this *Chapter*, realized by an addition of white noise into the dynamics. Explicitly, the oscillator in the (ξ_x, ξ_y)-state space reads

$$\frac{d}{dt}\begin{pmatrix} \xi_x \\ \xi_y \end{pmatrix} = \begin{pmatrix} 0 & 1 \\ -\omega_0^2 & 0 \end{pmatrix}\begin{pmatrix} \xi_x \\ \xi_y \end{pmatrix} + \begin{pmatrix} 0 \\ 1 \end{pmatrix} f(\xi_x, \xi_y) + \omega_0^2 \sqrt{2Q} \begin{pmatrix} 0 \\ 1 \end{pmatrix} \Gamma(t). \qquad (22)$$

---

[6] A Rayleigh oscillator $\ddot{x} + x - \dot{x} + \bar{\beta}\dot{x}^3$ also describes a van der Pol oscillator $\ddot{y} + y - \dot{y} + 3\bar{\beta} y^2 \dot{y}$ for the corresponding velocity $y = \dot{x}$.

Using the so-called van der Pol transformation, one can rescale time, basically to guarantee that the fundamental period of the oscillator equals 2π, and transform the oscillator into polar coordinates, that is, $\tau = \omega_0 t$ and $x = r\cos\theta$ and $y = -\omega_0 r\sin\theta$. By this the dynamics (22) reads

$$\frac{d}{d\tau}\begin{pmatrix}\xi_r\\\xi_\theta\end{pmatrix} = \begin{pmatrix}0\\1\end{pmatrix} - \frac{1}{\xi_r}\left\{\frac{f(r\cos\theta, -\omega_0 r\sin\theta)}{\omega_0^2} + \frac{\sqrt{2Q}}{\xi_r}\Gamma(\tau/\omega_0)\right\}\begin{pmatrix}\xi_r\sin\xi_\theta\\\cos\xi_\theta\end{pmatrix}. \quad (23)$$

After inserting the nonlinearities (21) one can average over a period 2π, which leads to a diffusion equation for amplitude r and phase θ (here = instantaneous frequency) in form of (see Daffertshofer, 1998, for more details)

$$\frac{d}{d\tau}p(r,\theta,\tau) \approx -\frac{\partial}{\partial r}\left[\left\{\bar{n}_0(r) + \frac{Q}{2r}\right\}p(r,\theta,\tau)\right] + \frac{Q}{2}\frac{\partial^2}{\partial r^2}p(r,\theta,\tau)$$
$$-\bar{\psi}_0(r)\frac{\partial}{\partial\theta}p(r,\theta,\tau) + \frac{Q}{2r^2}\frac{\partial^2}{\partial\theta^2}p(r,\theta,\tau), \quad (24)$$

with $\bar{f}_0(r) = -d\bar{V}_0/dr$ and

$$\bar{V}_0(r) = -\frac{1}{4}\left\{\alpha - \frac{\beta+\gamma}{8}r^2\right\}r^2, \text{ and } \bar{\psi}_0(r) := 1 + \frac{\eta}{8}r^2. \quad (25)$$

The oscillator is considered to evolve along a stable limit cycle, which implies that the amplitude's potential $\bar{V}_0(r)$ has a stable fixed point at a finite, non-vanishing value $r_0 = \pm 2\sqrt{\alpha/(\beta+\gamma)}$; see Fig. 2 (left panel) and the phase increases linearly in time yielding a frequency of $\bar{\psi}_0(r)$, i.e. the (corrected) frequency of a Duffing oscillator; if $\eta = 0$ the frequency equals that of a harmonic oscillator. Recall that time has been rescaled by ω_0.

Thus, the averaging results in a decoupling of amplitude and phase dynamics. In consequence, one finds for Eq. (23) a so-called stochastically equivalent system, i.e. a stochastic dynamics that yields an identical Fokker-Planck equation (24), in the form of

$$\frac{d}{d\tau}\begin{pmatrix}\xi_r\\\xi_\theta\end{pmatrix} = \begin{pmatrix}\bar{f}(\xi_r)\\\bar{\psi}(\xi_r)\end{pmatrix} + \frac{\sqrt{Q}}{\xi_r}\begin{pmatrix}\xi_r\Gamma^{(r)}_{\tau/\omega_0}\\\Gamma^{(\theta)}_{\tau/\omega_0}\end{pmatrix}. \quad (26)$$

In Eq. (26) the abbreviations (25) have been used as well as $\bar{f}(r) = -d\bar{V}/dr$ and

$$\bar{V}(r) := \bar{V}_0(r) - \frac{Q}{2}\ln r, \text{ and } \bar{\psi}(r) := \bar{\psi}_0(r) + \frac{Q}{2r^2}. \quad (27)$$

When comparing the forms (27) with the noiseless case (25), one can realize a diverging term $\propto \ln r$, which is added to the amplitude's potential $\bar{V}_0$. Of course, this merely reflects the negligible probability to find the oscillator at the origin: the origin is a fixed point but it is an unstable point; in the presence of noise it will never be occupied. However, this effect also yields a spurious drift in that the steady amplitude is increased, i.e. the effective potential $\bar{V}$ has minima at different locations

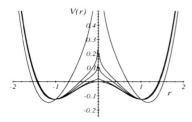 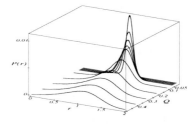

Fig. 2 Left panel: effective potential $\bar{V}$ of the amplitude dynamics for different noise strength Q, for definition see (27); right panel: corresponding stationary probability densities $p_{st} \propto \exp\{-\bar{V}/Q\}$; see text for further explanation.

than $\bar{V}_0$; see Fig. 2 and refer to Stratonovich (1963); Graham and Haken (1971). Also, the frequency of the oscillator is altered due to the presence of additive noise: it increases with increasing fluctuation strength Q but drops with increasing amplitude r. Put differently, the smaller the amplitude the faster the oscillation, an effect that has been ignored in more phenomenological studies on, e.g., additional noise in the (relative) phase dynamics of self-sustaining oscillators where the noise was just added without any dependency on the amplitude (e.g., Schöner et al, 1986).

3.5 Periodically Forced Potentials – Stochastic Resonance and More

Noise is usually considered detrimental for signal detection and information transmission, which appears obvious if the system under study is linear. In a linear system the response to the sum of multiple stimuli equals the sum of its responses to the corresponding individual stimuli, that is, (additive) noise can only reduce the signal-to-noise ratio when assessing the deterministic signal. If the system under study is nonlinear, the effects of adding noise may alter dramatically. For specific dynamical systems an increase in input noise can result in an increase in the signal-to-noise ratio (SNR), thereby enhancing the detectability of a signal. Benzi et al (1981) showed that a bistable system, subject to a periodic forcing in combination with a particular level of random perturbation may show a (local) maximum in is spectral distribution in the vicinity of the forcing frequency, which is absent when either the periodic forcing or the random perturbation have improper strength. This behavior is typically referred to as *stochastic resonance*, which has been mathematically discussed to great detail by McNamara and Wiesenfeld (1989). One may say that the signature of stochastic resonance is that the detection of a sub-threshold stimulus is optimally enhanced with a particular non-zero level of input noise, i.e. with increasing input noise signal-to-noise ratio increases to some peak and subsequently decreases.

Here, the phenomenon of stochastic resonance is illustrated using the example of the aforementioned bimodal potential in the presence of noise and a periodic external force; see Fig. 3. Explicitly, this bistable dynamics can read

$$\dot{\xi} = -\frac{dV}{d\xi} + A\sin\Omega t + \sqrt{2Q}\Gamma(t) \quad \text{with} \quad V(x) = -\frac{\varepsilon}{2}x^2 + \frac{1}{4}x^4 \qquad (28)$$

with a corresponding diffusion equation like

$$\dot{p}(x,t) = \frac{\partial}{\partial x}\left[\left\{\frac{dV}{d\xi} - A\sin\Omega t\right\}p(x,t)\right] + Q\frac{\partial^2}{\partial x^2}p(x,t). \qquad (29)$$

That is, the system response is driven by the combination of the two forces, i.e. the sinusoidal force and the noise, which compete/cooperate to make the system switch between the two stable states (located at the minimal of the bimodal potential). The switching rate thus depends on the interplay between the two forces and at a certain relationship a maximal response can occur in that the periodic driving is pronounced above the 'underlying' noise level. The point of this maximal signal-to-noise ratio *SNR* is hence a resonance induced by noise. The *SNR* of the dynamics (28) can be estimated as

$$SNR \propto Q^{-2}e^{-V_0/Q}. \qquad (30)$$

The potential relevance of stochastic resonance for motor control still needs to be explored. However, a process for which this interplay between noise and nonlinearity in a dynamical system might be of general importance is postural control as this is well-known to display many complex characteristic. Upright stance requires the (transient) stabilization of an unstable orbit; see *Paragraph 6*. Any small excursion from the equilibrium position during quiet stance may result in a progressive growth of the torque around the ankle due to gravity, accelerating the body away from the equilibrium position. To maintain upright stance, activity of muscles crossing the ankle joints is necessary to counteract the destabilizing torque due to gravity. Despite a huge variety of theories describing postural control, it is commonly accepted that the control relies on feedback mechanisms that are based on visual, vestibular, and proprioceptive information (see, .e.g., Peterka, 2002, for review). Part of the proprioceptive information is plantar tactile information. Meyer et al (2004) showed that, in the absence of visual information, postural sway velocity largely increased as a result of reduced plantar tactile sensation. If, by the process of stochastic resonance,

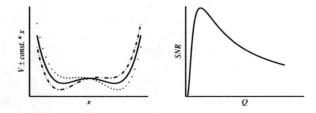

Fig. 3 Left panel: example of a bimodal potential (solid line) that is tilted (dashed and dashed-dotted lines) reflecting a periodic forcing. Right panel: Non-monotonic change in signal-to-noise ratio dependent on the noise strength. In a linear system one would find a monotonic decrease in the signal-to-noise ratio; see text for further details.

presenting noise to the mechanoreceptors improves the detectability of plantar tactile stimuli, this could facilitate postural control and therefore increase postural stability.

3.5.1 Periodic Potentials

In view of the afore-discussed HKB-model, which does not include a simple bimodal potential as in (28) as it has to apply for a 2π-periodic variable (the phase), one may ask if comparable resonance effects also occur in periodic potentials that are periodically forced. Thinking of so-called ratchet-dynamics, an asymmetry of the potential is the key ingredient yielding a net macroscopic current of ensemble member, i.e. all 'particles' move into one direction yielding a macroscopic drift. The HKB-potential, however, is symmetric. Do symmetric external forces influence stochastic systems when their nonlinear potentials are also symmetric? In fact, it has been shown that in systems like

$$\dot{\xi} = -\frac{dV}{d\xi} + A\sin\Omega t + \sqrt{2Q}\Gamma(t) \quad \text{with} \quad V(x) = -\cos x, \tag{31}$$

for which the Fokker-Planck equation is equivalent to Eq. (29) by substituting the according potential V, the diffusion rate can be greatly enhanced if the various forcings (periodic and noisy) are chosen in an optimal manner (Gang et al, 1996). In particular, one may obtain diffusion rates larger than the rate of free diffusion; see Fig. 4.

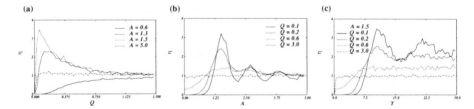

Fig. 4 Diffusion plots as function of noise strength and forcing. The figures show the ratio η between the free diffusion (diffusion in the absence of a potential, i.e. $V = 0$) and the diffusion of the system (31) for different parameter combinations: (a) η as a function of noise strength Q; (b) η as a function of forcing amplitude A; (c) η as a function of forcing period $T = 2\pi/\Omega$. In all cases an increase in diffusion rate as compared to the free diffusion can be observed, that is $\eta > 1$.

4 System Identification

While the analytical discussion in the previous sections served to illustrate in essence the qualitative effects of noise in nonlinear dynamical systems, the hands-on merits of this perspective for studying human movement have yet to be demonstrated. To this aim, a signal analysis approach to experimentally obtained data will be summarized and applied to finger tapping in order to extract its underlying stochastic dynamics.

Since its introduction by Friedrich and Peinke (1997), the extraction procedure has indeed found many applications in physics (e.g., Friedrich and Peinke, 1997; Waechter et al, 2003), engineering (Gradišek et al, 2000, 2002), economics (Friedrich and Peinke, 1997), sociology (Kriso et al, 2002), or meteorology (Sura, 2003), just to mention a few. The method has been successfully tested by analyzing physiological signals (Kuusela et al, 2003) and, here most importantly, kinematic data (van Mourik et al, 2006a, 2008; Huys et al, 2008; Gottschall et al, 2009). For instance, Frank et al (2006) derived a stochastic differential equation for isometric force production and showed that in their experimental data the force variability increased with the required force output because of a decrease of deterministic stability and an accompanying increase of noise intensity. Frank et al (2006) could determine a deterministic linear control loop and the random component involving a noise source that scales with force output. Before discussing these applications in the study of human movement in more detail, we outline the method's general principles and requirement as also outlined by van Mourik et al (2006a).

4.1 Drift and Diffusion Coefficients

The extraction of the deterministic and stochastic components is based on the calculation of probability densities. All the necessary numerical implementations to estimate probabilities and Kramers-Moyal coefficients are added (in form of Matlab functions) in *Appendix C*.

If we view human movement as a deterministic process with noise that obeys a system of stochastic differential equations, the time evolution of the corresponding probability density can be described by an equation of motion. Aim is thus to identify the drift and diffusion coefficients $D^{(1)}$ and $D^{(2)}$ in the Fokker-Planck equation (15) to reconstruct the dynamics (6) or (16). As explained above, the drift and diffusion coefficients are identical to the first and second-order cumulants or the first two Kramers-Moyal coefficients of the conditional probability density. In line with the previous form (14), a cumulant of an arbitrary order k (or the kth-order Kramers-Moyal coefficients) can be computed as

$$D^{(k)}(x) = \lim_{\Delta t \to 0} \frac{1}{k!} \frac{1}{\Delta t} \int [x' - x]^k p(x', t + \Delta t | x, t) \, dx', \qquad (32)$$

where Δt represents an infinitesimal time step as the limit approaches zero. The conditional probability density function $p(x', t + \Delta t | x, t)$ represents the probability of the system to be found in state x' at time $t + \Delta t$, given a previous state x at time t. Once this probability is estimated, Eq. (32) can be used to pinpoint drift and

Analyzing Noise in Dynamical Systems

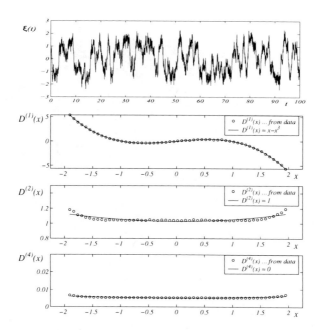

Fig. 5 Upper panel: A snapshot of a realization $\xi = \xi(t)$ of a stochastic system described by $\dot{\xi} = \varepsilon\xi - \xi^3 + \sqrt{2Q}\Gamma(t)$ with $Q = 1, \varepsilon = 1$, and time step 10^{-2}. Lower panels: Kramer-Moyal coefficient $D^{(1)}$, $D^{(2)}$, and $D(4)$. In particular the latter vanishes rendering the diffusion approximation valid; see text (for more details see also Van Mourik et al, 2006b).

diffusion coefficients[7]. First, the data have to be binned, that is, the range of values of each variable has to be subdivided into equally spaced parts or bins. Subsequently, the conditional probability density $p(x',t'|x,t)$ can be determined by computing the probability to find a sample at time t' in a bin with center x' assuming that at time t the previous sample was found in a bin with center x (note that $t' > t$). This computation has to be carried out for all neighboring pairs of samples and all combinations of bins. Then, according to Eq. (32), the resulting values of the conditional probability density are multiplied by their corresponding differences (raised to the power k) so that integration over the bins of the 'next' sample and scaling by the time step results in the proper drift and diffusion coefficients. Figure 5 illustrates the procedure in the one-dimensional case (see Van Mourik et al, 2006b, for more details).

Interestingly, this approach has been extended to weakly non-stationary data by which the analysis becomes applicable for systems exhibiting phase transitions.

---

[7] Before applying this extraction procedure, however, the underlying description in terms of stochastic dynamics given by equation (6) needs to be validated (see, e.g., van Mourik et al, 2006a, for more details). For instance, one has to verify whether the system under study can be described as a Markov process as mentioned earlier, that is, as a system whose future probability density depends only on its present value and not on its history; see *Paragraph 4.2*.

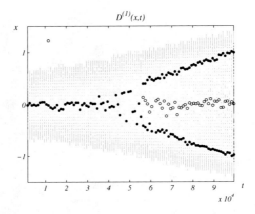

Fig. 6 Results of subsequent extractions of the drift coefficient of a Langevin system with cubic nonlinearity; cf. Fig. 5. 10^7 samples were simulated (time step 10^{-2}, $Q = 1/20$). Here we linearly increased ε from -1 to 1 resulting in a bifurcation at $\varepsilon = 0$. Per time window ($T = 10^5$ samples) we estimated $D^{(1)}$ and determined its extrema via a polynomial fit: filled dots represent stable points; open dots are unstable points.

To illustrate this 'moving window' technique in Fig. 6, we revisit the simulated, stochastic dynamics of Fig. 5 that includes a cubic nonlinearity and undergoing a super-critical pitchfork bifurcation via a slow increase of the corresponding bifurcation parameter.

4.2 Markov Properties and the Chapman-Kolmogorov Test

The sketched approach requires the system to be a Markov process. To test for Markov properties, we denote the system's state variables by x, and $p(x',t'|x,t)$ is the probability density to find the system at time t' at state x' where we presume the previous state x at time t (with $t'' > t' > t$). Then one has to verify the integral Chapman-Kolmogorov equation, which reads

$$p(x'',t''|x,t) = \underbrace{\int p(x'',t''|x',t') p(x',t'|x,t) dx'}_{= \tilde{p}(x'',t''|x,t)} \tag{33}$$

In words, one calculates the conditional probability for the time difference $t'' - t$ and compares it to the r.h.s. of (33) via, for instance, a conventional χ^2-statistics, as

$$\chi^2 = \iint \frac{[p(x'',t''|x,t) - \tilde{p}(x'',t''|x,t)]^2}{p(x'',t''|x,t) + \tilde{p}(x'',t''|x,t)} dx'' dx \tag{34}$$

If Eq. (33) holds, we have $p(x'',t''|x',t',x,t) = p(x'',t''|x,t)$, implying the 'absence of memory' in the system, i.e. Markovianity. Note, if this equality cannot be verified, then the outlined method is not applicable and may yield unpredictable results. In this case one should abstain from proceeding with this approach.

5 Rhythmic Movements

A major advantage of the just sketched analysis over conventional approaches is that the separation of the dynamics into Kramers-Moyal coefficients allows for detailed studies of smooth (differentiable) but otherwise arbitrary deterministic and stochastic parts in dynamical systems with noise. The method does not require any assumptions regarding possible analytical forms of the underlying, generating dynamics and can thus be viewed as an entirely unbiased tool. That is, the to-be-extracted dynamics is not limited to the aforementioned combination of Rayleigh and van der Pol oscillators, Duffing-like cubic or quintic stiffness, et cetera. Not only the steady-state but also transient behavior can be invoked, which may actually improve numerical estimates by increasing the phase space area that is accessible for analysis.

Fig. 7 depicts the method's implementation to synthetic tapping data. Areas in phase space can be identified and interpreted in terms of their respective experimental constraints, e.g., flexion/extension differences, 'discontinuities' like in tapping and anchoring phenomena. Notice that the suggested interpretation of vector fields (van Mourik et al, 2008) are by no means exhaustive; they merely served as easily accessible examples of the kind of information one might be able to glean from the extracted deterministic dynamics. Of course, this also applies to the interpretation of the stochastic component, which is not shown here but can be found in, e.g., van Mourik et al (2008). In combination with the study of local effects of deterministic forms this provides insight into the dynamical structure and the structure of noise as functions of location in phase space, also in relation to experimental conditions.

The extracted functions are not prescribed, i.e., no assumptions regarding appropriate analytical forms are required. If desired, however, analytical functions can be identified in the extracted dynamics, e.g., to quantitatively compare findings with earlier studies. Then, vector fields like the ones currently depicted can provide important means to constrain the modeler's intuition in choosing relevant analytical terms when seeking to reconstruct a certain dynamics. In (van Mourik et al, 2006a), particular dissipative terms could be identified as deterministic components of a limit cycle description of smooth rhythmic movements in which inertia and impact forces played a marginal role, whereas higher order terms and more dimensions appear to be indispensable in reconstructing the dynamical equations of motion for real tapping with contact.

The extraction procedure allows for a direct assessment and evaluation of the deterministic and stochastic parts of an experimental system of interest as represented by empirical data. Thus, the method may render an objective analysis tool that is, in principle, independent of a priori assumptions regarding the analytical form of the underlying dynamics. Possible changes of the dynamic structure due to altered experimental circumstances can be pinpointed by analyzing the extracted

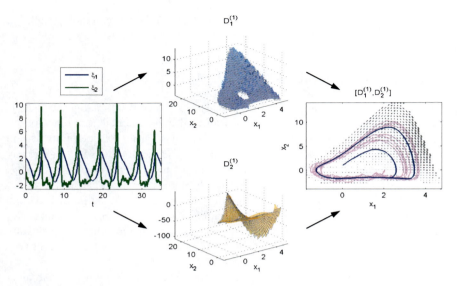

Fig. 7 Procedure to determine the dynamics of a time series, here tapping data. Position and velocity signals are simulated, ξ_1 and $\xi_2 = \dot{\xi}_1$, respectively. Kramers-Moyal coefficients $D_1^{(1)}$ and $D_2^{(1)}$ are determined using Eq. (14), with which the deterministic part of the dynamics can be reconstructed. Here a polynomial fit is added with which the deterministic dynamics can be written as $\dot{x}_1 \approx 0.2 + 0.2x_1 - 0.2x_1^2 - x_2 + 0.1x_1x_2$ and $\dot{x}_2 \approx -0.6 + 1.1x_1 + 0.2x_1^2 + 0.5x_1^3 - 0.7x_1x_2 + 1.2x_1^2x_2 - 0.6x_2^2$. The diffusion coefficient can be treated equally, see van Mourik et al (2008).

dynamics in its corresponding phase space. Interestingly, in a recent experiment in which human participants performed finger flexion-extension movements at various movement paces and under different instructions Huys et al (2008) used this approach for topological analyses of the flow in state space. In doing so it was shown that distinct control mechanisms underlie discrete and fast rhythmic movements: discrete movements require a time keeper, while fast rhythmic movements do not.

5.1 HKB-Bifurcation – Real Data

Van Mourik et al (2006b) applied aforementioned moving windows to experimental data, in which an external control parameter was slowly varied inducing a switch in the qualitative behavior of the steady response of the system under study ($\sim$ phase transition). Subjects produced isometric forces through thumb adduction in between the beats of a metronome. When the metronome's tempo was increased, subjects switched involuntarily to adduction at the beats of the metronome. I.e. there was a qualitative change in coordination from syncopation to synchronization.

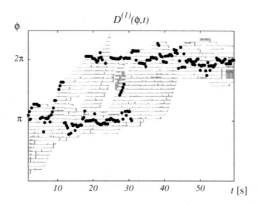

Fig. 8 Results of $D^{(1)}$ estimate using moving windows that were shifted along time series obtained in an experiment on coordinative instabilities (see text). The relative Hilbert phase $\phi(t)$ between metronome and response (produced force) was calculated from the data (sampling frequency 1250Hz). The change in metronome tempo from 1Hz to 3Hz induced a change in coordination from syncopation ($\phi \approx \pi$) to synchronization ($\phi \approx 2\pi$ or $\phi \approx 0$). The dots indicate stationary points that were determined via simulation of the reconstructed dynamics until steadiness.

The resulting bifurcation revealed Schöner's extension of the HKB-model (17) that here reads

$$\dot{\xi}_\phi = -\sin\xi_\phi - \frac{\varepsilon}{2}\sin 2\xi_\phi + \sqrt{2Q}\Gamma(t). \tag{35}$$

6 Posture – Center-of-Pressure Dynamics

The problem of postural balance is well-known for producing erratic motions of the center-of-mass and center-of-pressure (COP) alike. The complex interactions between the upright body and the environment result in quite irregular body sway patterns, which will here be analyzed by means of COP trajectories, the point location of the vertical ground reaction force vector, acquired from force platform data. The erratic fashion of COP migration is an indirect result of the complex interplay between gravity-line migration and inertial forces (Zatsiorsky & Duarte, 1999). Interestingly, previous research on COP time series analyses have repeatedly reported evidence for the existence of processes that take place on two different time scales (Amblard et al., 1985; Collins & De Luca, 1994; Frank et al., 2001; Zatsiorsky & Duarte, 2000). The mechanisms that account for such 'slow' and 'fast' dynamics have not yet been pinpointed explicitly but a number of authors have made a variety of suggestions.

Gottschall et al (2009) used the outlined extraction method to analyze the COP dynamics during quiet stance. Drift and diffusion coefficients were estimated and fitted by a simplified parametrization: a linear fixed point dynamics. While this approach appears interesting in that it capitalizes on the delicate interplay between the

Fig. 9 In (Boulet et al, 2010) visual feedback of the COP during standing was presented at delays ranging from zero to one second in increments of 250 milliseconds. Using stochastic delay differential equations, the COP and centre-of-mass (COM) dynamics could be explicitly modeled with two independent delay terms for vision and proprioception.

deterministic and stochastic components in the COP, one has to realize that more than a decade ago Collins and De Luca (1993); Collins and De Luca (1994) already analyzed COP trajectories as one- and two-dimensional random walks and extracted repeatable, physiologically meaningful parameters from stabilograms. Their posturographic analyses demonstrated that COP trajectories cannot be modeled as mere diffusion processes. Indeed the COP might be modeled via fractional Brownian motion and that at least two control systems – a short-term mechanism and a long-term mechanism – were operating during quiet standing. Fractional Brownian motion processes and according analysis methods are briefly summarized in *Appendices A & B*; Eke et al (see also 2002). One should realize that the COP dynamics may not fulfill the aforementioned Markov properties in *Paragraph 4.2* rendering the application of the present extraction procedure questionable. The non-Markov characteristics were recently accounted for by Boulet et al (2010) in a study on a visuo-postural control loop in upright stance; see Fig 9. Using stochastic differential equations, that by contrast to the afore-listed model included distinct delays, COP and centre-of-mass (COM) dynamics were explicitly modeled. Two independent delay terms were included, one for vision and another for proprioception. Using a novel 'drifting fixed point' hypothesis to describe the fluctuations of the COM with the COP being modeled as a faster, corrective process of the COM. The model turned out to agree well with the data in terms of probability density functions, power spectral densities, short and long-term correlations (i.e. Hurst exponents) as well as the critical time between the two ranges. The linearized dynamics of the COP along the anterior/posterior direction ξ_{AP} and the visual feedback ξ_{fb} reads as follows:

$$\begin{aligned}\dot{\xi}_{AP} &= \varepsilon \xi_{AP} - \kappa \xi_{AP}(t - \tau_{int}) + \sqrt{2Q_{int}}\Gamma^{(int)}(t) + \xi_{fb} \\ \dot{\xi}_{fb} &= -\gamma \xi_{AP}(t - \tau_{fb}) + \sqrt{2Q_{fb}}\Gamma^{(fb)}(t),\end{aligned} \qquad (36)$$

in which τ_{int} denotes an intrinsic delay (i.e., through delayed proprioception), τ_{fb} is the experimentally manipulated delay in the visual feedback, ε relates to the (linear) drift that causes the inverted pendulum 'body' to fall over, κ quantifies the corresponding restoring force, and γ parametrizes the tendency to put the visual feedback on target; see Fig. 9 and Boulet et al (2010) for detailed discussion. As said, the presence of delays introduces a non-trivial memory to the COP-dynamics. This kind of dynamics becomes more common in the study on human movement. Interestingly, Patanarapeelert et al (2006) investigated a linear stochastic delay differential equation of second order with time delay and computed its variance as a function of the time delay. Strikingly, close to the parameter regime in which their deterministic system exhibits Hopf bifurcations (switches between fixed-point to limit cycle dynamics), they found that the variance as a function of the time delay exhibits a sequence of pronounced peaks. These peaks were interpreted as delay-induced destabilization resonances arising from so-called oscillatory ghost instabilities. On the basis of the obtained theoretical findings, re-interpretations of previous human motor control studies and predictions for human motor control studies were provided. To what extent these approaches are indeed feasible remains to be seen. In any case, the lack of Markov properties forms a challenge when using mere time series for a stochastic system identification. A very promising approach to this important issue is the approximation of non-Markovian processes by so-called nonlinear Fokker-Planck equations (Frank, 2005), that is, diffusion equations like Eq. (15), but where the probability density p occurs in a nonlinear form (e.g., p^2). The extraction of drift and diffusion coefficients has indeed been extended to that case (Frank and Friedrich, 2005) but future work has to reveal the appropriateness of this method to the analysis of COP data.

7 Summary

Random noise is omnipresent in motor behavior. Any mathematical approach to the behavioral or neuronal dynamics alike hence requires a proper account for randomness. In this *Chapter*, the mathematical framework of stochastic differential equation has been briefly summarized. With this terminology at hand, some profound dynamical properties of stochastic systems were explained, for instance, spurious drifts or stochastic resonance. In addition to this 'forward modeling' approach, noise and its consequent statistics were used to define a generic analysis technique for extracting so-called drift and diffusion coefficients. This just recently established analysis method has been advocated as it allows for an unbiased identification of both types of system components. The deterministic components are revealed in terms of drift coefficients and vector fields (while the stochastic components can be assessed in terms of diffusion coefficients, see, e.g., van Mourik et al, 2008). The general principles of the procedure and its application were explained and exemplified by extracting deterministic and stochastic aspects of various instances of movement, including finger tapping and the erratic motion of the center-of-pressure. It was also indicated how the extracted numerical forms can be analyzed to gain insight into the dependence of dynamical properties on experimental conditions.

Acknowledgements. I would like to thank Jason Boulet, Mark Snaterse, Maarten van den Heuvel, Alistair Vardy, and in particular Anke van Mourik for providing substantial material for this summary. I also thank Ramesh Balasubramaniam, Peter Beek, Didier Delignières, André Longtin, and Melvyn Roerdink for the noisy discussions.

References

Beek, P.J., Peper, C.E., Daffertshofer, A.: Modeling rhythmic interlimb coordination: Beyond the Haken-Kelso-Bunz model. Brain and Cognition 48(1), 149–165 (2002)

Benzi, R., Sutera, A., Vulpiani, A.: The mechanism of stochastic resonance. Journal of Physics A 14(11), L453–L457 (1981)

Boulet, J., Balasubramaniam, R., Daffertshofer, A., Longtin, A.: Stochastic two delay-differential model of delayed visual feedback effects on postural dynamics. Philosophical Transactions of the Royal Society A 368(1911), 423–438 (2010)

Collins, J.J., De Luca, C.J.: Open-loop and closed-loop control of posture - a random-walk analysis of center-of-pressure-trajectories. Experimental Brain Research 95(2), 308–318 (1993)

Collins, J.J., De Luca, C.J.: Random walking during quiet standing. Physical Review Letters 73(5), 764–767 (1994)

Daffertshofer, A.: Effects of noise on the phase dynamics of nonlinear oscillators. Physical Review E 58(1), 327–338 (1998)

Eisenhammer, T., Hübler, A., Packard, N., Kelso, J.A.S.: Modeling experimental time series with ordinary differential equations. Biological Cybernetics 65(2), 107–112 (1991)

Eke, A., Herman, P., Kocsis, L., Kozak, L.: Fractal characterization of complexity in temporal physiological signals. Physiological Measurement 23(1), 1–38 (2002)

Frank, T., Friedrich, R.: Estimating the nonextensivity of systems from experimental data: a nonlinear diffusion equation approach. Physica A 347, 65–76 (2005)

Frank, T.D.: Nonlinear Fokker-Planck equations. Springer, Berlin (2005)

Frank, T.D., Friedrich, R., Beek, P.J.: Stochastic order parameter equation of isometric force production revealed by drift-diffusion estimates. Physical Review E 74(5), 051905 (2006)

Friedrich, R., Peinke, J.: Description of a turbulent cascade by a Fokker-Planck equation. Physical Review Letters 78(5), 863–866 (1997)

Gang, H., Daffertshofer, A., Haken, H.: Diffusion of periodically forced brownian particles moving in space-periodic potentials. Physical Review Letters 76(26), 4874–4877 (1996)

Gardiner, C.W.: Handbook of Stochastic Methods. Springer, Berlin (2004)

Gottschall, J., Peinke, J., Lippens, V., Nagel, V.: Exploring the dynamics of balance data - movement variability in terms of drift and diffusion. Physics Letters A 373(8-9), 811–816 (2009)

Gradišek, J., Siegert, S., Friedrich, R., Grabec, I.: Analysis of time series from stochastic processes. Physical Review E 62(3 A), 3146–3155 (2000)

Gradišek, J., Grabec, I., Siegert, S., Friedrich, R.: Qualitative and quantitative analysis of stochastic processes based on measured data, I: Theory and applications to synthetic data. Journal of Sound and Vibration 252(3), 545–562 (2002)

Graham, R., Haken, H.: Fluctuations and stability of stationary non-equilibrium systems in detailed balance. Zeitschrift für Physik 245, 141 (1971)

Haken, H.: Synergetics. Springer, Berlin (1974)

Haken, H., Kelso, J.A.S., Bunz, H.: A theoretical model of phase transitions in human hand movements. Biological Cybernetics 51(5), 347–356 (1985)

Harris, C.M., Wolpert, D.M.: Signal-dependent noise determines motor planning. Nature 394(6695), 780–784 (1998)

Has'minskiĭ, R.: Stochastic stability of differential equations. Sijthoff & Noordhoff, Rockville (1980)

Honerkamp, J.: Statistical Physics. Springer, Berlin (1998)

Hurst, H.: Long-term storage capacity of reservoirs. Transaction of the American Society for Civil Engineering 116, 770–799 (1951)

Huys, R., Studenka, B.E., Rheaume, N.L., Zelaznik, H.N., Jirsa, V.K.: Distinct timing mechanisms produce discrete and continuous movements. PLoS Computational Biology 4(4), e1000,061 (2008)

Kay, B.A.: The dimensionality of movement trajectories and the degrees of freedom problem: A tutorial. Human Movement Science 7(2-4), 343–364 (1988)

Kay, B.A., Kelso, J.A.S., Saltzman, E.L., Schöner, G.: Space-time behavior of single and bimanual rhythmical movements: Data and limit cycle model. Journal of Experimental Psychology: Human Perception and Performance 13(2), 178–192 (1987)

Kay, B.A., Saltzman, E.L., Kelso, J.A.S.: Steady-state and perturbed rhythmical movements: A dynamical analysis. Journal of Experimental Psychology: Human Perception and Performance 17(1), 183–197 (1991)

Kelso, J.A.S.: Phase transitions and critical behavior in human bimanual coordination. American Journal of Physiology - Regulatory Integrative and Comparative Physiology 15(6), R1000–R1004 (1984)

Körding, K.P., Wolpert, D.M.: Bayesian integration in sensorimotor learning. Nature 427(6971), 244–247 (2004)

Kramers, H.A.: Brownian motion in a field of force and the diffusion model of chemical reactions. Physica. 7(4), 284–304 (1940)

Kriso, S., Peinke, J., Friedrich, R., Wagner, P.: Reconstruction of dynamical equations for traffic flow. Physics Letters A 299(2-3), 287–291 (2002)

Kuusela, T., Shepherd, T., Hietarinta, J.: Stochastic model for heart-rate fluctuations. Physical Review E 67, (6, Part 1), 061,904 (2003)

Mandelbrot, B., van Ness, J.: Fractional Brownian motion, fractional noises and applications. SIAM Review 10(3), 422–437 (1968)

McNamara, B., Wiesenfeld, K.: Theory of stochastic resonance. Physical Review A 39(9), 4854–4869 (1989)

Meyer, P., Oddsson, L., De Luca, C.: The role of plantar cutaneous sensation in unperturbed stance. Experimental Brain Research 156(4), 505–512 (2004)

van Mourik, A.M., Daffertshofer, A., Beek, P.J.: Deterministic and stochastic features of rhythmic human movement. Biological Cybernetics 94(3), 233–244 (2006a)

van Mourik, A.M., Daffertshofer, A., Beek, P.J.: Estimating kramers-moyal coefficients in short and non-stationary data sets. Physics Letters A 351(1-2), 13 (2006b)

van Mourik, A.M., Daffertshofer, A., Beek, P.J.: Extracting global and local dynamics from the stochastics of rhythmic forearm movements. Journal of Motor Behavior 40(3), 214–231 (2008)

Moyal, J.E.: Stochastic processes and statistical physics. Journal of the Royal Statistical Society B 11, 150–210 (1949)

Patanarapeelert, K., Frank, T.D., Friedrich, R., Beek, P.J., Tang, I.M.: Theoretical analysis of destabilization resonances in time-delayed stochastic second-order dynamical systems and some implications for human motor control. Physical Review E 73(2), 021901 (2006)

Pawula, R.: Approximation of linear Boltzmann equation by Fokker-Planck equation. Physical Review 162(1), 186–188 (1967)

Peng, C., Havlin, S., Stanley, H., Goldberger, A.: Quantification of scaling exponents and crossover phenomena in nonstationarty heartbeat time-series. Chaos 5(1), 82–87 (1995)

Peterka, R.: Sensorimotor integration in human postural control. Journal of Neurophysiology 88(3), 1097–1118 (2002)

Post, A.A., Peper, C.E., Daffertshofer, A., Beek, P.J.: Relative phase dynamics in perturbed interlimb coordination: Stability and stochasticity. Biological Cybernetics 83(5), 443–459 (2000)

Rangarajan, G., Ding, M. (eds.): Processes with Long Range Correlations: Theory and Applications. Lecture Notes in Physics, vol. 621. Springer, New York (2003)

Riley, M.A., Turvey, M.T.: Variability and determinism in motor behavior. Journal of Motor Behavior 34(2), 99–125 (2002)

Risken, H.: The Fokker-Planck Equation. Springer, Berlin (1989)

Schöner, G.: A dynamic theory of coordination of discrete movement. Biological Cybernetics 63(4), 257–270 (1990)

Schöner, G.: Timing, clocks, and dynamical systems. Brain and Cognition 48(1), 31–51 (2002)

Schöner, G., Haken, H., Kelso, J.A.S.: A stochastic theory of phase transitions in human hand movement. Biological Cybernetics 53(4), 247–257 (1986)

Siefert, M., Kittel, A., Friedrich, R., Peinke, J.: On a quantitative method to analyze dynamical and measurement noise. Europhysics Letters 61(4), 466–472 (2003)

Sornette, D.: Critical Phenomena in Natural Sciences. Springer, Berlin (2004)

Stratonovich, R.L.: Topics in the Theory of Random Noise. Gordon and Breach, New York (1963)

Sura, P.: Stochastic analysis of southern and pacific ocean sea surface winds. Journal of the Atmospheric Sciences 60(4), 654–666 (2003)

Todorov, E., Jordan, M.I.: Optimal feedback control as a theory of motor coordination. Nature Neuroscience 5(11), 1226–1235 (2002)

Waechter, M., Riess, F., Kantz, H., Peinke, J.: Stochastic analysis of surface roughness. Europhysics Letters 64(5), 579–585 (2003)

Wax, M. (ed.): Selected papers on noise and stochastic processes. Dover, New York (1954)

Appendix A - More General, Continuous 1D Random Processes

A chapter on noise needs to address more formal definitions of general random processes, at least to some degree. Here we will briefly add processes that exhibit so-called power law characteristics. The mathematical forms go back to early studies of Mandelbrot and van Ness (1968), who searched for explanatory forms of what is these days called the Hurst effect: a nonlinear drop in the auto-correlation function of a process under study. The nonlinearity ought to reflect a scale-free (and self-similar) process so that the drop in correlation follows a power of the form $\propto t^{-2H}$ where H refers to the Hurst exponent.

A.1 Definitions

As usual in this *Chapter*, $\Gamma(t)$ denotes white noise with zero mean and unit variance. $\mathbb{E}[\xi]$ is the average over the random variable ξ in the sense of distributions[8]. The noise is considered 'white' because its frequency spectrum is flat, i.e. all frequencies

---

[8] By assuming ergodicity this average equals the mean over time but estimates over finite time spans should usually be seen as approximation.

are equally present. For the correlation this 'whiteness' translates into an 'uncorrelatedness' in the sense of a temporal δ-distribution, i.e. $\mathbb{E}\left[\Gamma(t)\Gamma(t')\right] = \delta(t-t')$.

More general random processes $\xi(t)$ can be considered as linear combinations of white noise, which in its easiest form may be realized as weighted sum or, given the continuous dependency of time, in the integral of time modulated by some delayed *kernel* function K, i.e. a K-convoluted or a K-filtered white noise. Starting at t_0 from an initial value ξ_0 this process hence may read

$$\xi(t) = \xi_0 + \varepsilon \int_{t_0}^{t} K(t-\tau)\Gamma(\tau)\,d\tau$$

The aforementioned properties of white noise yield the *average* of $\xi(t)$ as

$$\mu = \mathbb{E}\left[\xi(t)\right] = \mathbb{E}\left[\xi_0\right] + \varepsilon \int_{t_0}^{t} \mathbb{E}\left[\Gamma(\tau)\right] K(t-\tau)\,d\tau \stackrel{\mathbb{E}[\Gamma(\tau)]=0}{=} \mathbb{E}\left[\xi_0\right] = \mu_0$$

and the *auto-covariance* reads

$$\sigma^2(t,t') = \mathbb{E}\left[\left[\xi(t)-\mu\right]\left[\xi(t')-\mu\right]\right] = \sigma_0^2 + \varepsilon^2 \int_{t_0}^{\frac{1}{2}(t+t'-|t-t'|)} K(t'-\tau)K(t-\tau)\,d\tau$$

when σ_0^2 denotes the variance of the initial density $\{\xi_0\}$. Notice that we directly obtain the conventional, time-dependent variance by using $t' = t$ as

$$\sigma^2(t) = \sigma_0^2 + \varepsilon^2 \int_{t_0}^{t} K^2(t-\tau)\,d\tau$$

We further obtain the so-called *mean squared displacement* as

$$\psi^2(t,t') = \mathbb{E}\left[\left[\xi(t)-\xi(t')\right]^2\right] = \sigma^2(t) + \sigma^2(t') - 2\sigma^2(t,t')$$

Note that quite often the initial density $\{\xi_0\}$ is chosen as a mean-centered δ-distribution, i.e. $\mu_0 = 0$ and $\sigma_0^2 \to 0$.

A.2 Again the Wiener Process

Before analyzing more general forms let us briefly recapitulate the earlier discussed case of a Wiener process (11), for which

$$K(t-\tau) = \begin{cases} 1 & \text{for } 0 \leq \tau \leq t \\ 0 & \text{otherwise} \end{cases}$$

holds. That is, we have $\xi_w(t) = \xi_{0,w} + \varepsilon \int_0^t \Gamma(\tau)\,d\tau$. We replaced ξ by ξ_w for the sake of legibility. Using $\mu_0 = 0$, $\sigma_0^2 \to 0$, and $t_0 \to -\infty$ (or just $t_0 \leq 0$), the average, variance, etc. directly become $\mu_w = 0$, $\sigma_w^2(t) = \varepsilon^2 t$,

$$\psi_w^2(t,t') = \varepsilon^2 |t-t'| \quad \text{and} \quad \sigma_w^2(t,t') = \frac{1}{2}\varepsilon^2\left(t - |t-t'| + t'\right).$$

A.3 Fractional Brownian Motion

If the kernel is more complicated by means of

$$K(t-\tau) = \begin{cases} (t-\tau)^{H-\frac{1}{2}} & \text{for } 0 \leq \tau \leq t \\ (t-\tau)^{H-\frac{1}{2}} - (-\tau)^{H-\frac{1}{2}} & \text{otherwise} \end{cases}$$

and we directly use $\mu_0 = 0$, $\sigma_0^2 \to 0$, and $t_0 \to -\infty$, then the dynamics reads

$$\xi_{fBm}(t) = \varepsilon \int_{-\infty}^{0} \left[(t-\tau)^{H-\frac{1}{2}} - (-\tau)^{H-\frac{1}{2}}\right] \Gamma(\tau) d\tau + \varepsilon \int_{0}^{t} (t-\tau)^{H-\frac{1}{2}} \Gamma(\tau) d\tau.$$

Now the statistical measures[9] become $\mu_{fBm} = 0$ and $\sigma_{fBm}^2(t) = C\varepsilon^2 t^{2H}$. Furthermore we have

$$\psi_{fBm}^2(t,t') = C\varepsilon^2 |t-t'|^{2H} \quad \text{and} \quad \sigma_{fBm}^2(t,t') = \frac{1}{2}C\varepsilon^2 \left[t^{2H} - |t-t'|^{2H} + t'^{2H}\right].$$

A.4 Fractional Gaussian Noise

An often discussed case is that for discrete increments $t \to t + \Delta$ of the fractional Brownian motion as these increments turn out to be stationary. Since the increments' distribution is Gaussian they are referred to as fractional Gaussian noise $\xi_{fGn}(t)$ though we will not further pursue this notion. Instead we consider

$$\Delta \xi_{fBm}(t) = \xi_{fBm}(t+\Delta) - \xi_{fBm}(t) = \varepsilon \int_{t_0}^{t} K(t-\tau)\left[\Gamma(\tau+1) - \Gamma(\tau)\right] d\tau.$$

With $\mu_0 = 0$, $\sigma_0^2 \to 0$, and $t_0 \to -\infty$, we find $\mu_{\Delta fBm} = 0$ and

$$\psi_{fBm}^2(t+\Delta,t) = \sigma_{\Delta fBm}^2(t) = C\varepsilon^2 \Delta^{2H},$$

that is, the variance of the increments is independent of time t. The auto-covariance (here = auto-correlation) becomes

$$\sigma_{\Delta fBm}^2(t,t') = \frac{1}{2}C\varepsilon^2 \left[|t-t'+\Delta|^{2H} - 2|t-t'|^{2H} + |t-t'-\Delta|^{2H}\right].$$

and, in particular for $t' - t = T$ we have

$$\sigma_{\Delta fBm}^2(t,t+T) = \frac{1}{2}C\varepsilon^2 \left[|T-\Delta|^{2H} - 2|T|^{2H} + |T+\Delta|^{2H}\right],$$

which is also independent of time t and converges for $T \to \infty$ to

---

[9] Here we abbreviate with $C = 2\int_{0}^{\infty}\left[1 - \left(1+\frac{1}{\tau}\right)^{H-\frac{1}{2}}\right]\tau^{2H-1}d\tau$.

Analyzing Noise in Dynamical Systems

$$\lim_{T \to \infty} \sigma^2_{\Delta fBm}(t, t+T) = C\varepsilon^2 \frac{H\Delta^2}{2H-1} T^{2H-2}.$$

For $\frac{1}{2} < H < 1$ the auto-covariance or auto-correlation decays so slowly that its integral diverges. Put differently, for $\frac{1}{2} < H < 1$ correlations in the fractional Gaussian noise are persistent, whereas for $0 < H < \frac{1}{2}$ correlations are anti-persistent.

Fractional Gaussian noise is sometimes referred to as 'derivative' of fractional Brownian motion, i.e. $\tilde{\xi}_{fGn}(t) = \dot{\xi}_{fBm}(t)$, with

$$\dot{\xi}_{fBm}(t) = \varepsilon \int_{t_0}^{t} \frac{dK(t-\tau)}{dt} \Gamma(\tau) d\tau + \varepsilon \lim_{\tau \to t} K(t-\tau) \Gamma(t)$$

$$= \varepsilon \int_{-\infty}^{t} (t-\tau)^{H-\frac{3}{2}} \Gamma(\tau) d\tau + \ldots$$

There, previous calculations for the fractional Brownian motion can be, by and large, adopted[10] after substituting H by $H - 1$.

Appendix B - A Bit on Time Series Analysis

In following, estimates of power law characteristics and scaling exponents H are briefly sketched to provide a link with linear time series analysis. An in-depth discussion of the methods is, however, beyond the scope of the current *Chapter*. For the interested reader (Rangarajan and Ding, 2003; Sornette, 2004) provide very useful entry points into this exciting topic.

B.1 *Power Spectra: The Wiener-Khinchin Theorem*

One can relate the auto-covariance function to the spectral density, a fact that is often exploited to estimate the correlation characteristics in time series, e.g., power laws, so-called one-over-f features, et cetera. The Wiener-Khinchin theorem states

$$P_{fBm}(\omega) = \mathbb{E}\left[\xi^*_{fBm}(\omega) \xi_{fBm}(\omega)\right] \propto \frac{1}{|\omega|^{2H+1}} = |\omega|^{-\beta}$$

where $\xi_{fBm}(\omega)$ denotes the time series ξ_{fBm} Fourier transform and '*' refers to the conjugate complex value. This implies in all generality $\beta = 1 + 2H$. For example, conventional Brownian motion ($H = \frac{1}{2}$) has $\beta = 2$. White noise, by contrast has, $\beta = 0$ implying, in turn, $H = -\frac{1}{2}$, i.e. fractional Gaussian noise as derivative of fractional Brownian motion with $H = \frac{1}{2}$. The case $\beta = 1$ corresponds to $H = 0$. Note that we also have

---

[10] Notice that with $\dot{x}_{fBm}(t) = \varepsilon \int_{-\infty}^{t} (t-\tau)^{\vartheta - 1} \Gamma(\tau) d\tau + \ldots$ one often finds that $0 < \vartheta < \frac{1}{2}$ corresponds to persistent correlations in the fractional Gaussian noise while $-\frac{1}{2} < \vartheta < 0$ implies that correlations are anti-persistent.

$$P_{\Delta fBm}(\omega) = \mathbb{E}\left[\xi^*_{\Delta fBm}(\omega)\xi_{\Delta fBm}(\omega)\right] \propto \frac{1}{|\omega|^{2H-1}} = |\omega|^{-\tilde{\beta}}.$$

B.2 (Rescaled Range) R/S-Statistics: Hurst Effect

In his original paper Hurst (1951) studied the rescaled range that is defined as

$$\left.\begin{array}{l} R_\xi(t,\Delta) = [\sup\limits_{0\leq\delta\leq\Delta} - \inf]\left\{\int_t^{t+\delta}\xi(\tau)d\tau - \frac{\delta}{\Delta}\int_t^{t+\Delta}\xi(\tau)d\tau\right\} \\ S_\xi(t,\Delta) = \sqrt{\frac{1}{\Delta}\int_t^{t+\Delta}\left[\xi(\tau) - \frac{1}{\Delta}\int_t^{t+\Delta}\xi(\tau')d\tau'\right]^2 d\tau} \end{array}\right\} \to (R/S)_\xi = \frac{R_\xi(t,\Delta)}{S_\xi(t,\Delta)};$$

Here, sup and inf refer to the limit superior and limit inferior, respectively. Exploiting the self-similarity properties, Mandelbrot and van Ness (1968) have shown that for the discrete increments of a fractional Brownian motion the following power law holds

$$\frac{R_{fBm}(t,\Delta)}{S_{fBm}(t,\Delta)} \propto \Delta^{H+1} \quad \text{and} \quad \frac{R_{\Delta fBm}(t,\Delta)}{S_{\Delta fBm}(t,\Delta)} \propto \Delta^H.$$

Hurst found in his data the H is typically larger than $\frac{1}{2}$, i.e. the underlying fractional Gaussian noise has persistent (long-term) correlations.

B.3 Detrended Fluctuation Analysis

Instead of the aforementioned R/S-statistics one can alternatively write $\tilde{\xi}(\Delta;t) = \int_t^{t+\Delta}\xi(\tau)d\tau$ and define a linear trend $m_\Delta \delta$ with $m_\Delta(t) = \frac{1}{\Delta}\tilde{\xi}(\Delta;t)$, which leads to

$$R_{\tilde{\xi}}(t,\Delta) = [\sup\limits_{0\leq\delta\leq\Delta} - \inf]\left\{\tilde{\xi}(\delta;t) - m_\Delta(t)\delta\right\}.$$

One may further improve this guessed linear trend $m_\Delta(t)$ with by an estimate of its mean over the time span $[t,\ldots,t+\Delta]$, yielding time-dependent coefficients $a_\Delta(t)$ and $b_\Delta(t)$ that minimize $\mathbb{E}\left[\left[\tilde{\xi}(\Delta;t) - a_\Delta(t)t - b_\Delta(t)\right]^2\right]$. If this trend is subsequently removed[11], the remainder is analyzed in terms of its variance or standard-deviation (Peng et al, 1995): $DFA_\xi(\Delta) = \sqrt{\sigma_{\hat{y}}^2(\Delta)}$ with $\hat{y}(\Delta;t) = \int_t^{t+\Delta}\xi(\tau)d\tau - [a_\Delta(t)t + b_\Delta(t)]$. If ξ represents the discrete increments of a fractional Brownian motion, i.e. $\xi(t) = \Delta\xi_{fBm}(t)$, then its integral displays fractional Brownian motion and we find

$$DFA_{fBm}(\Delta) \propto \sqrt{\Delta^{2H+2}} = \Delta^{H+1} \quad \text{and} \quad DFA_{\Delta fBm}(\Delta) \propto \sqrt{\Delta^{2H}} = \Delta^H.$$

---

[11] One may also say that per interval $[t,\ldots,t+\Delta]$ eventual offsets in $\xi(t)$ are eliminated piecewise.

Appendix C - Matlab Codes

The Matlab functions below can be used as starting point to implement the extraction method summarized in *Paragraph 4*. This section merely provides some ideas of the numerics. Source codes, including graphical output and a full documentation can be accessed via http://www.move.vu.nl/members/andreas-daffertshofer.

C.1 Examples

One-dimensional case (Fig. 5); use the dynamics

$$\dot{\xi} = \varepsilon\xi - \xi^3 + \sqrt{2Q}\Gamma(t)$$

and sample 10^6 data points at a rate of 100Hz ($\varepsilon = 1, Q = 1$), compute $D^{(1)}$ and $D^{(2)}$ for 50 bins and plot the results as function of the state-space x:

```
xi=pitchfork_sde(1,1,10^6,1/100,0);
[D,x]=KM(xi,50,1/100,[1,2],1,100);

i=find(isfinite(D{1}) & isfinite(D{2}));
X=x{1}(i);

for k=1:2
    subplot(2,1,k);
    plot(X,D{k}(i),'bo');
    c=polyfit(X,D{k}(i),3*(k==1));
    hold on; plot(X,polyval(c,X),'k-'); hold off;
end
```

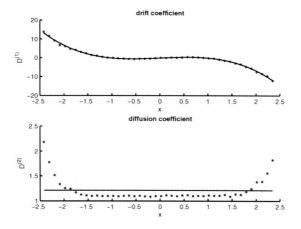

Two-dimensional case, cf. Eqs. (21 & 22); use the oscillator dynamics

$$\ddot{\xi} = -\kappa_1\xi + \kappa_2\dot{\xi} + \kappa_3\dot{\xi}^3 + \kappa_4\xi^2\dot{\xi} + \kappa_5\xi^3 + \sqrt{2Q}\Gamma(t)$$

and sample 10^6 data points at a rate of 100Hz ($\kappa = \{1, 0.5, 0, 1, 1\}, Q = 0.5$), compute $D_1^{(1)}$ and $D_2^{(1)}$ for 50 bins and plot the resulting 2-dim. vector field:

```
xi=osci_sde([1,1,0,-1,0],0.5,10^6,1/100,0);
[D,x]=KM(xi,50,1/100,[1],1,100);

[X,Y]=ndgrid(x{1},x{2});
i=find(isfinite(D{1})&isfinite(D{2}));

plot(xi(1:5000,1),xi(1:5000,2));
hold on;
quiver(X(i),Y(i),D{1}(i),D{2}(i));
hold off;
```

Three-dimensional case, the Lorenz attractor; use the dynamics

$$\dot{\xi} = \kappa_1 (\psi - \xi) + \sqrt{2Q}\Gamma_\xi(t)$$
$$\dot{\psi} = \xi (\kappa_2 - \chi) - \psi + \sqrt{2Q}\Gamma_\psi(t)$$
$$\dot{\chi} = \xi\psi - \kappa_3\chi + \sqrt{2Q}\Gamma_\chi(t)$$

and sample 10^6 data points at a rate of 100Hz ($\kappa = \{10, 28, 8/31\}, Q = 5$), compute $D_{1\ldots 3}^{(1)}$ for 50 bins and plot the resulting three-dim. vector field:

```
xi=lorenz_sde([10,28,8/3],1,10^6,1/100,0);
[D,x]=KM(xi,50,1/100,[1]);

[X,Y,Z]=ndgrid(x{1},x{2},x{3});
i=isfinite(D{1}) &...
  isfinite(D{2}) & ...
  isfinite(D{3});

plot3(xi(:,1),xi(:,2),xi(:,3),...
  'color',[.7,.7,.7]);
hold on;
quiver3(X(i),Y(i),Z(i),...
  D{1}(i),D{2}(i),D{3}(i));
hold off;
```

Analyzing Noise in Dynamical Systems

C.2 The Core Function KM.m

Computation of n-dimensional Krames-Moyal coefficients; for an extended version of KM.m consult the aforementioned URL:

```
function [D,x,orders]=KM(xi,bins,dt,order,transitionStep,minNumOfVals)
% compute bin edges bins and centers x
[x,bins]=bincenters(xi,bins);
[len,dim]=size(xi);
% define bin arrays
M=ones(1,dim); for k=1:dim, M(k)=length(x{k}); end
numbins=prod(M);
if numbins>2^32, error('total # of bins is too large (max = 232)'); end
% assign every sample in xi a integer index uniquely defining a bin
binnedvalue=int32(ones(1,len));
for d=1:dim
    index=false(len,1);
    nb=prod(M(1:d-1));
    for k=1:M(d)
        i=xi(:,d)<=bins{d}(k+1);
        index=xor(index,i);
        binnedvalue(index)=binnedvalue(index)+nb*(k-1);
        index=i;
    end
end
% determine number of (unique) KM coefficients for a given order ...
nKM=0; for o=1:length(order), nKM=nKM+nPolyExp(dim,order(o)); end
% ... and initialize output arrays D ...
D=cell(nKM,1); for d=1:nKM, D{d}=NaN*zeros([M,1]); end
% ... and output orders
orders=cell(nKM,1); j=1; for o=1:length(order)
    n=sortrows(polyExp(dim,order(o)),-1);
    for k=1:size(n,1), orders{j}=n(k,:); j=j+1; end
end
% and loop along all bins ...
for k=1:numbins
    index=binnedvalue==k; % get the number of entries in xi at bin k
    s_index=sum(index);
    if s_index>minNumOfVals % if a min. number of entries is present
        % compute the conditional histgram: take only the already found
        % entries, go steps ahead (transitionStep)
        p=reshape(histc(binnedvalue([false(1,transitionStep),...
            index(1:end-transitionStep)]),1:numbins),[M,1]);
        % normalize histrogram to obtain the probability distribution
        n=p; for d=1:dim, n=trapz(x{d},n,d); end; p=p/n;
        A=ind2subC(dim,[M,1],k);
        % compute the KM coefficients
        for o=1:nKM % loop over all dimensions and orders
            aux=p;
            for d=1:dim
                if orders{o}(d)
                    dist=(x{d}-x{d}(A(d))).^(orders{o}(d));
                    m_ind=mod(fix((0:numbins-1)/prod(M(1:d-1))),M(d));
                    for j=1:M(d)
                        ind=find(m_ind==j-1);
                        aux(ind)=dist(j)*aux(ind);
                    end
                end
            end
            for d=1:dim, aux=trapz(x{d},aux,d); end
            D{o}(k)=(1/factorial(sum(orders{o})))*aux/dt;
        end
    end
end
```

C.3 The Simulated Models

A one-dimensional dynamics with cubic nonlinearity:

```
function x=pitchfork_sde(epsilon,Q,N,dt,x0)
langevinforce=sqrt(2*Q*dt)*randn(N,1); % define the Langevin force
x=zeros(N,1); x(1)=x0;                 % initialize the output array
for k=2:N                              % and Euler-forward integration
    x(k)=(1+epsilon*dt)*x(k-1)-x(k-1)^3*dt+langevinforce(k);
end
```

A two-dimensional oscillator containing Rayleigh, van der Pol, and Duffing terms:

```
function x=osci_sde(kappa,Q,N,dt,x0)
langevinforce=sqrt(2*Q*dt)*randn(N,1); % define the Langevin force
x=zeros(N,2); x(1,:)=[x0(:)]';         % initialize the output array
for k=1:N-1                            % Euler-forward integration
    x(k+1,1)=x(k,1)+x(k,2)*dt;
    x(k+1,2)=x(k,2)+(-kappa(1)*x(k,1)+kappa(2)*x(k,2) ...
        +kappa(3)*x(k,2).^3+kappa(4)*x(k,1).^2.*x(k,2) ...
        +kappa(5).*x(k,1).^3)*dt+langevinforce(k);
end
```

The Lorenz system with three noise terms:

```
function x=lorenz_sde(kappa,Q,N,dt,x0)
langevinforce=sqrt(2*Q*dt)*randn(N,3); % define the Langevin force
x=zeros(N,3); x(1,:)=[x0(:)]';         % initialize the output array
for k=1:N-1                            % and Euler-forward integration
    x(k+1,1)=x(k,1)+kappa(1)*(x(k,2)-x(k,1))*dt+langevinforce(k,1);
    x(k+1,2)=x(k,2)+((kappa(2)-x(k,3))*x(k,1)-x(k,2))*dt+langevinforce(k,2);
    x(k+1,3)=x(k,3)-(kappa(3)*x(k,3)-x(k,1)*x(k,2))*dt+langevinforce(k,3);
end
```

C.4 Some Auxilliary Functions

```
function [x,b]=bincenters(xi,b) % Definition of bins given a data array
[len,dim]=size(xi);
if iscell(b)==0, bb=b; b=cell(size(bb,2),1); for k=1:size(bb,2), b{k}=bb(:,k); end, end
if length(b)==1, b(1:dim)=b;
elseif length(b)~=dim, error('xi-dim and b-dim must match');
end
x=cell(dim,1); for k=1:dim
    if length(b{k})==1, b{k}=binset(min(xi(:,k)),max(xi(:,k)),b{k}); end
    x(k)=(b{k}(1:end-1)+b{k}(2:end))/2; b{k}(end)=Inf;
end

function b=binset(xmin,xmax,N)
b=(xmin:(xmax-xmin)/N:xmax)';

function c=ind2subC(nout,siz,ndx) % determine the subscript values
% corresponding to a given single index into an array:
if length(siz)<=nout, siz=[siz ones(1,nout-length(siz))];
else siz=[siz(1:nout-1) prod(siz(nout:end))];
end
k=[1 cumprod(siz(1:end-1))]; c=zeros(1,length(siz));
for i=length(siz):-1:1,
    vi=rem(ndx-1,k(i))+1; vj=(ndx-vi)/k(i)+1; c(i)=vj; ndx=vi;
end

function num=nPolyExp(dim,order) % counting elements in n-dim. polynomials:
num=nchoosek(dim+order,order)-nchoosek(dim+order-1,order-1);

function exps=polyExp(dim,order)
if dim==1, exps=order; return; end
if order==0, exps=zeros(1,dim); return;
elseif order==1, exps=eye(dim); return;
end
num=npolyExp(dim,order); exps=zeros(num,dim);
j=0; for k=0:order
    e=polynomExponents(dim-1,order-k);
    exps(j+(1:size(e,1)),1)=k; exps(j+(1:size(e,1)),2:end)=e;
    j=j+size(e,1);
end
```

The Dynamical Organization of Limb Movements

Raoul Huys

Abstract. The early 1980s saw the development of a new perspective on motor control inspired by theories of self-organization and dynamical systems theory. Its first efforts were directed at the investigation of rhythmic movements in terms of two-dimensional (autonomous) limit cycle oscillators. The corresponding studies are characterized by the development of detailed and generally task-specific models, which have resulted in a detailed documentation of the relation between oscillator properties and task requirements. The study of discrete movements for a long time received far less attention; its conjunctional theoretical and empirical investigation has only recently set off and is characterized by an explicit focus on phase flows and topologies therein.

1 Introduction

Over the last 25 years or so, the notion that (perceptual) motor behavior may be conceived of in terms of dynamical structures, such as limit cycles and fixed points, has become widely accepted. Why so? What do we learn about the control of movements under this perspective? Do we actually learn anything about motor control or is the corresponding research essentially a sophisticated descriptive data fitting exercise (cf. Rosenbaum, 1998)? What has instigated the notion that by definition abstract mathematical structures are an appropriate conceptualization of motor behavioral patterns that originate from a biological structure in the first place? These and related questions are central to the present chapter with a focus on single end-effector movements (for the coordination between movements, see the chapter of Calvin and Jirsa, this volume). While the present chapter only scratches the surface of the wealth of studies that have focused on this issue, we hope that it will provide an overview communicating the gist of the approach.

Single limb movements are oftentimes categorized as discrete, continuous or rhythmic. Discrete movements constitute singularly occurring motion preceded

---

Raoul Huys
Theoretical Neuroscience Group, Université de la Méditerranée, UMR 6233
"Movement Science Institute", CNRS, Faculté des Sciences du Sport,
13288, Marseille cedex 09, France

and followed by a period without motion (i.e., with zero velocity and acceleration). Continuous movements lack such recognizable start and end points, and are referred to as rhythmic if a particular 'movement event' is periodic (that is, repeats itself with a period T). Rhythmic movements have, in the dynamical framework, received far more attention than discrete movements, which follows from theoretical, methodological, and historical reasons. As for the latter, the development and popularization of the dynamical paradigm owes much to the study of the coordination between rhythmic movements and their formalization by Haken, Kelso and Bunz (1985; the HKB-model). A crucial aspect here is that the coordination between two limbs in terms of coupled oscillators allows for the derivation of relative phase between them (see Calvin & Jirsa, this volume). The relative phase follows a one-dimensional dynamics that allows for an easily comprehensible conception in terms of a potential landscape, rendering the formulation mathematically more accessible. Indeed, the larger part of all studies inspired by the HKB-model focuses solely on the relative phase dynamics and ignores how the individual oscillators contribute to this dynamics. Methodologically, rhythmic movements have the advantage of being repetitive, which allows for the collection of considerably longer time series than is the case for non-rhythmic movements. Consequently, a battery of time series analysis techniques can be used that are either inaccessible to discrete movements or, at least, will have less statistical power. Finally, rhythmic movements are theoretically attractive as they, at least in principle, allow for mathematically autonomous description, which is not the case for non-rhythmic movements. The previous two issues have been essential for the widespread study and maturation of the dynamic approach; the latter one, however, played a central role in its early development.

In the following, we will discuss the investigation of rhythmic movements followed by a discussion of their discrete counterparts. In that regard, while few (if any) would deny that biological systems are inherently stochastic, the present chapter focuses solely on their deterministic part (see the chapter of Daffertshofer, this volume, for the issue of stochasticity). We will first, however, briefly outline the theoretical motivations that underlie the conceptualization of motor control in terms of self-organized dynamical structures.

2 The Birth of the Dynamical Perspective

The influence of the Russian physiologist Nikolai Aleksandrovich Bernstein (1896/1966) on motor control theory can hardly be overestimated. Bernstein (in the 1967s English translation) posited the 'problem' of control and coordination in the course of action as a degrees-of-freedom problem. The human motor apparatus, for instance, comprises more than 200 bones, 110 joints and over 600 muscles, each one of which either spans one, two or even three joints. While the degrees of freedom are already vast on the biomechanical level of description, their number becomes dazzling when going into neural space. Functional goal-directed behavior requires that a certain order arises in this multi-degree of freedom system. From a control-theoretical perspective, this poses a seemingly unsolvable problem. Bernstein's gist was that during action, these degrees of freedom are temporally organized into a functional unit, referred to as a synergy or coordinative structure, so that the (mechanical) degrees of freedom are effectively minimized. But how are synergies formed? What principles underlie their formation?

These were the questions that Kugler, Kelso and Turvey addressed in their 1980 seminal paper, in which the conceptualization of movement and control in terms of dynamical structures, in particular rhythmic movements in terms of nonlinear oscillators (i.e., limit cycles), became firmly established. At the time, the main stream conception of motor control was coined in terms of central representations and 'higher level algorithmic computations' (by necessity involving an intelligent regulator) imposing control onto the 'lower level machinery' (the skeletal-muscular system). This notion basically boils down to the idea that an organism's sensory organs allow it to construct a representation of the environment on the basis of which it may compute a set of appropriate motor commands that are sent to the muscular apparatus (see for instance Richard Schmidt's widely adhered to famous general motor program; Schmidt, 1975; Schmidt & Lee, 2005). A few scholars, however, expressed their dissatisfaction with this essentially Cartesian paradigm (cf. Meijer & Roth, 1988, for discussions between opponents and proponents of this view), which stems from two insights. First, the implication of representations implies an entity, a controller or goal-directed agent that understands and acts upon them. In other words, it implies positing an entity similar to the one that is to be explained in the first place. The question to be answered is merely shifted from one entity (or level) to the next, inevitably leading to infinite regress (Dennett, 1978; 1991). Second, the embodiment of the computing entity is problematic—computing automata deal with quite different constraints than biological organisms: while the former deal with mathematical and logical constraints, the latter are inherently subjected to physical and biological ones.

Bernstein's degree-of-freedom problem (1967) may be viewed as a particular case of a more general question that was discussed in the 1970s among theoretical biologists, namely how order arises in biological systems (cf. Pattee 1972, 1973; Iberall 1970, 1977, 1978; and in particular a series edited by Waddington, 1968-1972). At the same time, theories of self-organization, the emergence and transitions between spatiotemporal patterns were developed in theoretical physics (Haken, 1977; Nicolis & Prigogine, 1989; Prigogine 1969; Prigogine & Nicolis, 1977). Accordingly, order and disorder, and transitions between ordered states may arise in open systems (i.e., systems with a continuous flux of energy, matter and/or information) that are far from thermodynamic equilibrium. These systems consist of numerous non-linearly interacting elements (Haken, 1977, 1983, 1996) and are referred to as 'dissipative structures' (Prigogine & Nicolis, 1977), which do not drift toward thermodynamic equilibrium, but rather conserve their stability through energy dissipation.

Inspired by these developments, and asserting that repetitive cycles of events are ubiquitous to biological systems (see Goodwin, 1963; 1970; Iberall, 1970, 1977, 1978), Kugler et al. pointed out that stability should be understood in a dynamical sense (1980, p 15; see also Chapter 1 by Fuchs, this volume). Indeed, under the premise that cyclicity is a manifestation of a universal design principle for autonomous systems (cf. Kelso, Holt, Kugler & Turvey, 1980, and Kelso, Holt,

Rubin, & Kugler, 1981, and references therein), the focus on oscillatory phenomena (rhythmic movements) imposes itself naturally. When using ordinary differential equations, a general dynamical formulation of an oscillatory system reads

$$\ddot{x} + f(x,\dot{x}) + g(x) = F(t) \tag{1}$$

where $\dot{x}$ and $\ddot{x}$ represent the first and second time derivatives of x, $f(x,\dot{x})$ the damping (or dissipation) function, $g(x)$ the stiffness (or elasticity) function, and $F(t)$ a time-dependent forcing function. (Inertia equals $m\ddot{x}$, m being mass, which we set here to 1.) The continuation of cyclical events can come about either through a forcing function $F(t)$ (in which case the system is non-autonomous due to its time dependency) or by the proper regulation of energy inflow and outflow based on the oscillatory motion itself (i.e., $F(t) = 0$, in which case the system is autonomous). A non-autonomous formulation was not sought for, as it necessitates explanation of the forcing function $F(t)$, that is, its inclusion puts the causation on another, not further explained level and thus paves the way to infinite regress (see also above). Oscillatory motion is self-sustaining (autonomous) if the energy inflow and outflow depend on the oscillator properties only and cancel each other out over a cycle. This is the case for so-called limit cycle oscillators, for which the energy book keeping depends on the damping (or dissipation) function. How energy is lost and inserted in the cycle depends on the specifics of the oscillator, that is, its stiffness and damping terms. In the well-known van der Pol oscillator, for instance, the damping function $f(x,\dot{x})$ equals $-\dot{x}(1-x^2)$, and is negative for $|x| < 1$, in which case energy is inserted into the system while for $|x| > 1$ energy is lost. Van der Pol damping is thus position dependent. In contrast, the Rayleigh damping function (i.e., $f(x,\dot{x}) = -\dot{x}(1-\dot{x}^2)$) is velocity dependent. The energy book keeping in these limit cycles thus depends on the position versus velocity of the system's motion, respectively. The identification of an oscillator's 'ingredients' thus reveals, among others, the variable that determines the energy injection into the system and where this occurs.

In view of these considerations, Kugler et al. (1980) proposed to conceptualize coordinative structures, which they defined as "a group of muscles often spanning a number of joints that is constrained to act as a single functional unit" (p. 17), as dissipative structures. Its stable state is maintained via energy freed up by metabolic processes. It should be noted that synergies are spatiotemporal organizations; the constituting components are temporally assembled so as to form a functional collective. (For example, the muscles spanning the ankles, knees and hips which are coordinated in a particular manner during walking are also involved in hopping albeit in a different manner.) The conceptualization of coordinative structures, or synergies, in terms of limit cycles, or dissipative structures, constitutes a non-reductionistic approach, since it does not attempt to describe behavior in terms of the (lower) level of biomechanical or neurophysiological processes or mechanisms. The damping and stiffness terms do not simply map onto its counterparts in biomechanical variables. The behavior of the system in equation 1 above can be fully described by the so-called state variables x and $\dot{x}$. In general, the variables that capture the system's order at the macroscopic level chosen are referred to as collective variables or order parameters (in synergetics). While these variables in general are abstract, they are thought to reflect the system's underlying (neural — at least in the context of human and animal motion) organization in a specific context[1].

3 The Empirical Study of Rhythmic Movements as Limit Cycles

The properties of limit cycles are well established in dynamical systems theory (cf. Guckenheimer & Holmes, 1983; Perko, 1991; Strogatz, 1994; Jordan & Smith, 1999). Fundamental, in that regard, is the notion that a limit cycle is an isolated closed trajectory (in phase or state space) and is stable (or attractive) if all neighboring trajectories approach it (see also Fuchs, chapter one, this volume). By implication, a perturbed trajectory should return to the limit cycle. This line of reasoning was adopted by Kelso et al. (1981) to investigate if the conception of rhythmic human movement in terms of limit cycles cuts ice. In their experiments, participants were instructed to perform cyclical finger flexion–extension movements at a self-chosen, comfortable pace. At specific moments, the moving finger was mechanically perturbed. It was found that the oscillation frequency and amplitude following the perturbation was not different than prior to it. In other words, the cyclical movements revealed a certain orbital stability, which was the first confirmation (to our best knowledge) for the portrayal of rhythmic movements in terms of limit cycles.

A detailed quantitative investigation to identify the oscillator underlying human movement was picked up by Kay and colleagues (1987) by examining the effect of oscillation frequency on the corresponding amplitude. In their study, the participants executed amplitude-unconstrained wrist movements at preferred frequency as well as at frequencies from 1 to 6 Hz (with 1 Hz steps). With increasing frequency, movement amplitude decreased while peak velocity increased. In addition, for each frequency a strong linear correlation was found between movement amplitude and peak velocity. Following Haken et al. (1985), and using averaging techniques from oscillator theory (the slowly varying amplitude and rotating wave approximation; cf. Strogatz, 1994; Jordan, & Smith, 1999) to study amplitude and peak velocity as a function of frequency, Kay et al. established that their participants' wrist movements could be accurately modeled by the hybrid oscillator combining van der Pol ($\sim \gamma x^2 \dot{x}$) and Rayleigh damping ($\sim \beta \dot{x}^3$);

---

[1] Fundamental to synergetics are the circular causality and slaving principle, which formulate the relation between the behavior of a system's multi-element microscopic level and the macroscopic behavior. While the latter reflects the behavior of the ensemble of microscopic elements, theirs is, in turn, enslaved by the macroscopic dynamics (cf. Haken, 1983; 1996). This bi-directional mechanism is formalized in a material substrate-independent manner.

$$\ddot{x} + \alpha\dot{x} + \beta\dot{x}^3 + \gamma x^2\dot{x} + \omega^2 x = 0 \tag{2}$$

Across participants, the experimentally observed frequency amplitude and frequency–velocity relations could be realized by adjustment of the linear stiffness parameter ω. These results further testified to the existence of limit cycles in (human) motor behavior.

Subsequently, Kay, Saltzman and Kelso (1991) investigated whether human rhythmic movements were autonomous and exhibit equifinality, as should be the case under their conception in terms of limit cycles. A system possesses equifinality if it reaches its equilibrium independent of initial conditions (implying relaxation back onto the limit cycle following perturbations). In the experiment, the participants were instructed to rhythmically move their index finger at their preferred frequency. In perturbation trials, mechanical perturbations in two directions and of two different magnitudes were delivered at eight equidistant positions in the movement cycle. Movement frequency, amplitude and peak velocity were examined prior to and following the perturbations — these variables should return to their pre-perturbation values following perturbations under the assumption that the movements are governed by a limit cycle dynamics. In addition, to estimate the strength of the attractor Kay et al. examined the relaxation time (i.e., the time taken to return to the limit cycle, which is shorter the stronger the attractor) as well as the phase response to further characterize the limit cycle (cf. Winfree, 1980). For instance, simple sinusoidally forced linear damped mass-spring systems reveal no phase shift (after a transient) following perturbation, while the amount of phase shift of different autonomous limit cycles depends on the magnitude and position of the perturbation. It was found that the frequency, amplitude and peak velocity were similar before and after the perturbations and were all independent of the perturbation properties. Furthermore, the relaxation time scaled with the magnitude of the perturbation. These observations are in line with the hybrid oscillator model proposed in the previous study (Kay et al., 1987). Two observations, however, were not compatible with it. First, the attractor's strength (assessed via the exponential return of the system back onto the limit cycle), but not relaxation time, appeared non-uniform in the phase plane. Second, a phase-dependent phase advance was observed; on average, the movement was temporarily sped up directly following the perturbation. To account for these observations Kay et al. suggested that either the autonomy assumption may have been incorrect — the system under study may have been forced or that an oscillator at the neural level driving a peripheral oscillatory component may be a better conception of human rhythmic motion than a two-dimensional autonomous oscillator.

An alternative explanation was brought forward by Beek and colleagues (Beek, Rikkert & van Wieringen, 1996). These authors set out to verify the hybrid model and to examine whether the dynamical features are limb independent — after all, even though the description is abstract in the sense that the coordination dynamics cannot be trivially reduced to biomechanical and neurophysiological factors, the dynamics still arise within a particular biological substrate. These considerations motivated the choice to investigate elbow movements, rather than wrist movements, while further remaining as close as possible to the study of Kay et al.

(1987); similar movement frequencies were implemented and the same kinematic relationships were examined. In contrast to Kay et al. however, Beek and colleagues performed the analysis on individual performances so as to assess how and to which degree individual (anatomical and neurophysiological) differences were expressed in the observable dynamics. Beek et al. found that whereas the (individually based) observed frequency–amplitude relations were consistent with the hybrid model, the frequency–peak velocity relationships were not. This inconsistency could be solved, however, by making the Rayleigh damping frequency dependent (i.e., $\beta \dot{x}^3$ in equation 2 becomes $\beta \omega \dot{x}^3$). Consequently, the stiffness and damping terms are linked, and the energy book keeping appears as a function of frequency. Further, the preferred frequency of wrist rotations in Kay et al. appeared about twice as high as the elbow rotation, which supposedly reflected inertial differences between the respective oscillators. The comparison across these studies highlighted that even though the abstract dynamics could not be trivially reduced to the underlying biological structure, it clearly was molded by it. In fact, a similar influence was found to be exerted by various task constraints, in that the composition of the damping (and stiffness) function depended on the task requirements under which a coordinative structure was assembled (Beek, Schmidt, Morris, Sim & Turvey, 1995).

While the above may seem to suggest that the oscillatory properties of human movement are limited to van der Pol and Rayleigh forms, other forms are possible as well. As early as 1988, Beek and Beek derived a 'catalog' of nonlinear oscillator ingredients (i.e., the terms appearing in $f(x,\dot{x})$ and $g(x)$ in equation 1 above) and discussed a variety of graphical techniques allowing for their identification from measured data (Beek & Beek, 1988). The authors distinguished conservative terms, that is, those that do not change the limit cycle's total energy and non-conservative terms that do affect the limit cycle's energy booking (see Table 1). The former conservative terms tend to influence the oscillation

Table 1 Admissible series expansion of $f(x,\dot{x})$ and $g(x)$. Adopted from Beek & Beek (1988).

Conservative: $f(x,\dot{x})=$	$x^0, x^2, x^4, \ldots$	van der Pol series
	$\dot{x}^0, \dot{x}^2, \dot{x}^4, \ldots$	Rayleigh series
	$x^0\dot{x}^0, x^2\dot{x}^2, x^4\dot{x}^4, \ldots$	π-mixed series (even)
Non-conservative: $f(x,\dot{x})=$	$x\dot{x}, x\dot{x}^3, \ldots$	
	$x^3\dot{x}, x^3\dot{x}^3, \ldots$	π-mixed series (odd)
$g(x)=$	$x^3, x^5, x^7, \ldots$	Duffing series

frequency but not its amplitude, and will thus play a role when precise timing is required. The non-conservative terms, in contrast, do not affect frequency but play a role when spatial precision is required, as in reaching movements.

4 Rhythmic Movements under Precision Requirements

In the studies discussed so far, no explicit demands were placed on the movements' amplitude. Quite often, however, movements are executed so as to manipulate the environment, such as reaching for an object. In such cases, movement accuracy determines whether or not the goal underlying the action will be reached. It is well known that movement accuracy and amplitude are inversely related to movement time (Fitts & Petterson, 1964; Meyer, Abrams, Kornblum, Wright, & Smith, 1988; Woodworth, 1899). The most well-known formulation of this so-called speed–accuracy trade-off was formalized by Fitts (1954). Accordingly, movement time MT equals $a + b \times ID$, where a and b represent parameters to be experimentally determined, and ID – the index of difficulty, a measure reflecting task difficulty, relates to target distance D and width W according to $ID = \log_2(2D/W)$. That is, movement time increases linearly with the ID via an increase of distance D and/or a decrease of width W. To examine how distance and accuracy constraints influence the dynamics in a repetitive precision aiming task (Fitts' task), Mottet and Bootsma (1999) examined Fitts' task from the perspective that rhythmic precision aiming is governed by limit cycle dynamics. In their experiment, participants performed a rhythmic aiming task at 18 IDs, ranging from 3 to almost 7 by manipulating distance and accuracy. The data were, next to phase plane representations, represented in (normalized) Hooke's portraits (i.e., the space spanning position versus acceleration). This representation was chosen to (in particularly, locally) assess the stiffness function (Guiard, 1993, 1997). In this representation, a purely linear (harmonic) oscillator appears as a straight line; deviations thereof reflect the influence of nonlinearities. The data clearly showed that the participants moved in an almost pure harmonic fashion at low IDs (indicated as a straight line in the Hooke's portrait), and that the influence of nonlinearities increased with increasing ID (the straight line became more and more N-shaped). Mottet and Bootsma (see also Bootsma, Boulard, Fernandez & Mottet, 2002: Mottet & Bootsma, 2001) proposed a minimal dynamical model to account for these main features including Rayleigh damping and Duffing stiffness,

$$\ddot{x} + c_{10}x - c_{30}x^3 - c_{01}\dot{x} + c_{03}\dot{x}^3 = 0 \tag{3}$$

The authors associated the stiffness with the time constraints inherent to the task and the damping with its spatial constraints (see also above). Mottet and Bootsma showed that through the appropriate parameterization of the damping and stiffness terms, their model was able to capture most of the variance (albeit decreasingly so with increasing ID). The model falls short, however, in that it cannot account for cases in which the stiffness locally vanishes (i.e., when the position reaches the root of the stiffness function), which would imply that the system would diverge to infinity (i.e., target overshoots cannot be dealt with). This problem can potentially be overcome, however, by including an additional stiffness term (a quintic Duffing term; cf. Mottet & Bootsma, 1999; Schöner, 1990). Furthermore, their data indicated that at low *ID*s (below approximately 4), the dissipative Rayleigh term was negative, rendering the limit cycle unstable: a flaw that can be overcome by adding van der Pol terms, and assuming these to be stronger than the Rayleigh terms at low *ID*s and weaker than them at high *ID*s.

As the task difficulty increases in a Fitts' task, the movements not only become slower and slower but corrective sub-movements also start to occur during the final 'homing-in' phase (at the target; cf. Elliot, Helsen, & Chua, 2001; Meyer et al., 1988). In addition, the time on the target (the dwell time) also increases (Adam & Paas, 1996; Buchanan, Park, Ryu, & Shea, 2003). That is, the movements obtain discrete characteristics. During discrete movements, energy is totally dissipated when the oscillation is at its excursions, i.e., the ratio of the acceleration at its maximal absolute position and the maximal acceleration vanishes. In contrast, harmonic movements reveal total energy conservation and the ratio above equals one (cf. Guiard, 1993, 1997). In the performance of a repetitive Fitts' task from low to high ID, this ratio goes from approximately one to zero. In other words, the (periodic) movements are continuous under low accuracy constraints and become discrete when the accuracy constraints are stringent, which led Mottet and Bootsma (1999) to suggest that the Rayleigh–Duffing model may provide a single dynamical structure allowing for rhythmic and discrete movements under a range of accuracy constraints.

A potential point of critique of the studies discussed so far is that the tasks therein lacked ecological validity, and one may question whether the concepts and methods that hold in highly constrained laboratory settings are also valid under ecologically more valid contexts, that is, in more complex tasks. It appears so. Beek (1989; see also Beek & Beek, 1988) investigated expert jugglers' hand motions during juggling. Using graphical and statistical analysis, he found clear indications that the part of the cycle where the hand carried a ball was governed by van der Pol behavior, while Duffing behavior was observed when the hand was unloaded (i.e., from the toss to the catch). In other words, the juggling hand cycle contains two dynamical régimes. In addition, he reported a structural involvement of discrete forcing 'pulses' at the position of the point of ball release (the throw) and, to a lesser extent, the point of the catch. Beek concluded that while an autonomous description of a juggler's hand was a good first order approximation, a detailed description required the incorporation of discrete pulses. Beek suggested that the location of forced pulses most likely depended on the phasing of the airborne balls rather than on the dynamics of the hand movement itself. Rather than proposing that the forcing is time-dependent, he suggested that the forcing is informational (i.e., a function of information instead of time per se). A fully autonomous description of the system should thus in this instance not be sought at the level of the hand movement's description solely, but requires the incorporation of an information-driven forcing.

5 Perspectives Incorporating Discrete Movements

In the introduction, we provided a kinematically based definition of discrete and continuous movements — rhythmic or otherwise — in terms of the presence versus absence of a motionless period preceding and following a movement. While this definition is correct (taking into consideration that the first two time derivatives of position vanish; cf. Hogan & Sternad, 2007), it remains silent as regards the control structure underlying the movements. In that regard, there has been a longstanding debate in the literature on whether motor control is fundamentally discrete (in which case rhythmic movements are mere concatenations of discrete motion elements) or rhythmic (in which case discrete movements are merely aborted rhythmical movements) or whether both movements are controlled distinctly and cannot be reduced to each other (cf. Huys et al., 2008; Sternad, 2008).

Kugler et al. (1980)[2], in the paper briefly discussed above (see also the accompanying paper of Kelso et al., 1980) argued against the need to conceive of discrete and rhythmic behaviors as arising from different mechanisms. These authors indicated that a mass-spring system (see equation 1) may, via the appropriate parameterization of damping and stiffness, reveal discrete behavior by moving to an equilibrium point in the absence of oscillations. Since mass-spring systems are intrinsically rhythmic, discrete and rhythmic movements may be conceived of as different manifestations of the same organization, as argued by the authors.

It took, however, another 10 years before a full-fledged, explicit dynamical model was developed by Schöner (1990) in his attempt to provide a unified dynamical model to account for posture (the absence of movement), as well as rhythmic and discrete movements. To that aim, Schöner used the Piro-Gonzales oscillator (Gonzales & Piro, 1987),

$$\dot{x} = v$$
$$\dot{v} = -(a^2 + \omega^2)x + 2av - 4bx^2v + 2abx^3 - b^2x^5 \qquad (4)$$

(here, we neglect the noise term incorporated in Schöner, 1990). This particular structure was chosen as it is analytically solvable and allows for several dynamical régimes including a single fixed point, two simultaneously stable fixed points and a limit cycle — the dynamical ingredients necessary to model posture, and discrete and rhythmic movements. In addition, Schöner used the concept of behavioral information, which he and Kelso had previously introduced to account for modifications of dynamic patterns via behavioral requirements (such as environmental, intentional, learned and so on) in terms of the dynamics (Schöner & Kelso, 1988; see also Kelso, 1995). In this conception, the behavioral information (directly) contributes to the phase flow (or vector field; see below) in an additive fashion,

---

[2] In essence, Kugler et al. discussed Feldman's equilibrium point model (Feldman, 1980, 1986) in terms of dynamical systems.

and the intention to move (but not the movement's temporal evolution) is treated as such. After incorporation of the additive contribution of intention, the equation reads

$$\dot{x} = v$$
$$\dot{v} = -f(x,v) + \chi_{[t_0,t_0+\Delta t]}(t) c_{int}(t) f_{int}(x,v)$$
(5)

where χ indicates the time interval $[t_0,t_0+\Delta t]$ during which the intention to move is 'on' (i.e., it is equal to 1 within the interval and 0 outside it), and the strength of the behavioral information c_{int} is larger than zero. The f_{int} takes the system from the fixed point(s) régime to the limit cycle régime (its simplest form reads $f_{int} = -x$). Intention thus stabilizes the limit cycle, and a discrete movement can be seen as a limiting case of a rhythmic movement (i.e., it constitutes a half or a full cycle). It should be noted that in Schöner's conception discrete — but not rhythmic — movement execution requires that the phase flow underwriting the movement changes on the time scale of the movement.

6 Phase Flows and Topologies

From the sections above, it is clear that much effort has been directed into the explicit characterization of rhythmic motor behavior through the identification of damping and stiffness terms and thereby learn about the oscillator's properties under particular task realizations. A similar degree of explicitness characterizes Schöner's model (1990), even though the development of his rationale for choosing the particular model formulation appears (partly) vector field based. In other words, model development was driven by the aim to examine quantitative features. In contrast, Jirsa and Kelso (2005) recently formulated a model construct with the aim to provide a general theoretical framework of human movement and coordination — its scope is therefore qualitative, but allowing for quantitative precision if so required. Jirsa and Kelso's perspective explicitly stresses phase flow properties and, in particular, the topological structure(s) therein. The importance of the concept of phase (or state) space is that deterministic, time-continuous and autonomous systems can be unambiguously described through their flow in it, which in the case of movement is the space spanned by the system's position and velocity: at least so under the commonly adopted assumption that movements allow for description in two dimensions (but see below). While the phase flow quantitatively describes the system's evolution as a function of its current state, its qualitative behavior is uniquely determined by its topology. For the classification of dynamical systems, two theorems are of key importance. First, for two-dimensional systems, the Poincaré–Bendixson theorem states that if a phase plane trajectory is confined to a closed bounded region that contains no fixed points, it must eventually approach a closed orbit. By implication, chaos cannot occur in two-dimensional systems and the possible dynamics in these systems are severely limited: only fixed points, limit cycles and separatrices can exist (see also Fuchs, chapter one, this volume). Second, the Hartman–Grobman theorem states that the phase flow near a hyperbolic fixed point[3] is topologically equivalent to the phase flow of the linearization. This theorem implies the existence of a continuous invertible mapping, a so-called homeomorphism, between both local phase flows. While the Hartman–Grobman theorem is solely valid for fixed points, the notion of topological equivalence applies to other topological structures as well. Intuitively, one may think of dynamical systems as being topological equivalent if their underlying structure remains invariant under particular 'distortions'. For instance, 'bending' and 'stretching' (of closed orbits, for instance) are allowed, but disconnecting closed trajectories is not (cf. Steward, 1995, chapter 10 and 12, for an intuitive, non-technical introduction). The upshot is that only a limited number of dynamical structures can be found in two-dimensional systems, and that dynamical systems belong to the same class if, and only if, they are topologically equivalent. In other words, armed with theorems from dynamical system theory, it is possible to uniquely define movements classes. Phase flow topologies identify all behavioral possibilities within a class: while all behaviors within a class can be mapped upon others, such maps do not exist between classes (see also Huys et al., 2008b).

The model construct proposed by Jirsa and Kelso (2005) that allows for fixed points and limit cycles in phase space (so to account for discrete and rhythmic movements, respectively) reads

$$\dot{x} = \left[x + y - g_1(x) \right] \tau$$
$$\dot{y} = -\left[x - a + g_2(x,y) - I \right]/\tau \qquad (6)$$

Here, $\dot{x}$ and $\dot{y}$ represent the time derivatives of x and y, respectively, τ a time constant and in which I represents an external (instantaneous) input. For an appropriate choice of $g_1(x)$ and $g_2(x,y)$, this model belongs to the class of excitable systems (cf. FitzHugh, 1961; Murray, 1993). Furthermore, $g_1(x)$ and $g_2(x,y)$ have to be chosen so as to assure the system's boundedness and to implement specific task constraints, respectively. The minimal realization satisfying these (and certain other) constraints for $g_1(x)$ and $g_2(x,y)$ are $g_1(x) = x^3/3$ and $g_2(x,y) = b \cdot y$ (see Jirsa & Kelso for details). Under this realization, the parameters a and b determine whether the topological structures in phase space are one or two fixed points and a separatrix (a structure that locally divides the phase space in regions with opposing flows) or a limit cycle. The presence or absence of particular topological structures can be readily illustrated on the basis of the system's nullclines, that is, the curves in phase space where the flow is purely horizontal (i.e., $\dot{y} = 0$) or vertical (i.e., $\dot{x} = 0$). The nullclines in the present realization are defined by

$$\left. \begin{array}{l} \dot{x} = 0 = x + y - x^3/3 \\ \dot{y} = 0 = -x + a - by \end{array} \right\} \Rightarrow \begin{cases} y = -x + x^3/3 \\ y = (a-x)/b \end{cases} \qquad (7)$$

---

[3] A fixed point of an nth-order system is hyperbolic if all eigenvalues λ of the linearization have a non-vanishing real part (i.e., $\Re(\lambda_i) \neq 0$ for $i = 1...n$), which is the case for most fixed points (and those under consideration at present).

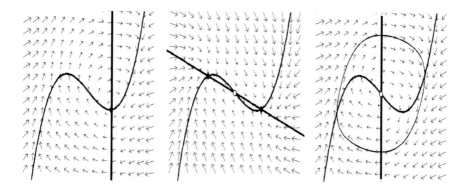

Fig. 1 Nullclines in phase space in the mono-stable (middle panel; $a=1.05$, $b=0$), bi-stable (left panel; $a=0$, $b=2$) and limit cycle condition (right panel; $a=0$, $b=0$). The dark gray arrows indicate the flow in phase space; the black lines represent the nullclines. Black dots represent stable fixed points; white dots represent unstable fixed points. The gray closed orbit in the right panel represents a stable limit cycle. No closed orbits exist in the mono- and bi-stable conditions.

(setting $I = 0$ and $\tau = 1$). A fixed point is found at the intersection of both nullclines. From these equations and Figure 1 it can be appreciated that only the nullcline with no vertical flow is affected by the parameterization and that its translation (via a) or rotation (via b) determines whether we find a single stable fixed point (referred to as the mono-stable condition), two stable fixed points interspersed with an unstable one (referred to as the bi-stable condition) or a stable limit cycle.

Two aspects of this perspective are crucial. First, with an eye on Figure 1 it can be readily appreciated that the phase space topology will be invariant under multiple choices for $g_1(x)$ and $g_2(x,y)$: numerous different nullclines can yield the particular intersections found in Figure 1 (under the adoption of some constraints; see Jirsa & Kelso [2005] for a detailed discussion). To reiterate, while different choices of $g_1(x)$ and $g_2(x,y)$ will affect detailed features of the phase flow, and thus the resulting quantitative behavior, the system's qualitative behavior is independent hereof. In this sense, this perspective may be said to be model independent. Second, a system settled at a fixed point will by definition stay there for all time unless an external 'force' kicks it away from it across the separatrix (see also Figure 1). That is, in the mono- and bi-stable condition the system will traverse through phase space (i.e., execute a movement) only in the presence of an external input I. In the case of a voluntary self-paced action, the external input I can hypothetically be considered as the activity of some neural (timer) structure. The system is thus non-autonomous (and strictly speaking [at least] 3-dimensional). In the limit cycle case, however, the system is autonomous and no input is required.

For discrete movement initiation, the separatrix can be thought of as a threshold mechanism: only if the system is brought across, a movement will be executed. Its existence (in the mono- and bi-stable conditions) predicts that the system may execute false starts, that is, execute a movement in the absence of an external stimulus (see Figure 2). Demonstrating the occurrence of false starts in humans would thus provide evidence that humans utilize this dynamic control mechanism. In the absence of an 'external' stimulus, noise could bring the system across the separatrix, although the noise strength thereto will under natural conditions most likely (and luckily!) be too small for this to occur. Fink and colleagues (Fink, Kelso & Jirsa, 2009) recently found evidence that humans utilize the mono-stable mechanism by evoking false starts via the application of mechanical perturbations. If a perturbation brings the systems nearer to the separatrix, then the additional impact of noise will be more likely to get the system across the separatrix. In the experiment, participants were instructed to execute a finger flexion–extension movement as fast as possible upon the presentation of an auditory stimulus (i.e., a reaction time task). In a quarter of the trials, a mechanical perturbation was applied either in the movement direction (i.e., flexion) or opposed to it (i.e., extension) just before the auditory stimulus presentation. In the trials without a perturbation, false starts were observed only in 2% of the trails. In contrast, false starts occurred in 34 and 9% of the trails in which perturbations were applied in the motion direction or opposed to it, respectively. This finding, and in particular the perturbation–direction dependence on the percentage of false starts, supports the idea that a motor control mechanism involving a separatrix is used by humans.

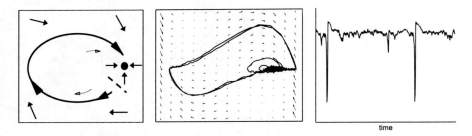

Fig. 2 The fixed point and separatrix in the mono-stable condition. The left panel represent the caricature of the phase flow in the mono-stable condition; the arrows indicate the flow, the black dot the fixed point, and the separatrix is visualized as the dashed line. A movement will be executed only if the system is brought across the separatrix (be it by an external stimulus or noise). The middle panel represents the phase space and its flow, and a trajectory of the model simulations without input I but with noise added of equation 7 (for a = 1.05, b = 0, and τ = 1, after transformation of the state variables into position and velocity, see Jirsa & Kelso, 2005). The left panel represents the trajectory in the middle panel as a time series. Here, two movements were executed as noise brought to the system across the separatrix.

The perspective outlined above inspired Huys and colleagues (Huys et al., 2008) to examine the implementation of discrete and rhythmic movements as a function of movement frequency. To investigate the behavioral capacity of the system described in equation 6 in the mono-stable and limit cycle régime, they computationally analyzed it under a wide range of parameters (including frequency ω). In addition, Huys et al. investigated human unimanual finger movements in comparable movement frequency régimes. In the experiment, the participants executed auditory-paced flexion–extension finger movements at frequencies from 0.5 Hz up to 3.5 Hz (with steps of 0.5 Hz) under the instruction to move as fast as possible, as smooth as possible or without any specific instruction (referred to as the 'fast', 'smooth', and 'natural' condition, respectively). The simulated data and human data were both examined using several measures; the behavioral data were also subjected to a vector field reconstruction (the Kramers-Moyal [KM] expansion; see Daffertshofer, chapter two, this volume and references therein). The computational analysis indicated that at low frequencies, the 'discrete' system (i.e., the model prepared in the mono-stable régime) produced trajectories similar to those of the human participants under the 'natural' and 'fast' conditions at low frequencies. At higher frequencies, the pacing of the external stimuli interfered with the system's 'intrinsic dynamics' (which occurs due to the occurrence of the stimulus before the system reaches the fixed point) and the required movement frequency could no longer be achieved. In contrast, in the limit cycle régime, the system was able to comply with all imposed temporal demands. It thus appeared that an externally driven system *cannot* produce movements at high frequencies while satisfying the required temporal constraints.

As for the human data, the reconstruction of the vector fields underlying the behavioral data were clear cut: under the 'fast' condition a fixed point was identified in the phase flows at low frequencies (and for half of the participants in the 'natural' condition). At high frequencies, however, irrespective of the instructions, no indications for the existence of fixed points were found, and it was concluded that the participants' movements were thus governed by limit cycle dynamics (which could not have been otherwise under the assumption that the phase space for such movements is two dimensional). This conclusion was supported by the other analyses. These results were taken to indicate that humans can employ two different timing mechanisms and naturally switch from a discrete mechanism to a rhythmic mechanism at a certain frequency, and that by implication, a movement initiation and timing mechanism should be involved at low, but not at high, frequencies.

In a follow-up study, Huys, Fernandez, Bootsma and Jirsa (2010) asked how spatial constraints impact the implementation of the identified timing mechanisms in the context of the well-known speed–accuracy trade-off (Woodworth, 1899) using Fitts' task (Fitts, 1954; Fitts & Peterson, 1964 see also above). Ten participants repetitively moved a stylus between two targets for 30 cycles (per trial) under instructions stressing speed as well as accuracy. Ten different target widths were used resulting in ten *ID*s ranging from 2.1 to 6.9. The data analysis was based on the phase plane analysis as above. It was found that at low *ID*s, the participants' movements were governed by limit cycle dynamics, while at high *ID*s they were governed by fixed point dynamics. The transition occurred at an *ID* of around 5. In other words, when the task difficulty increases humans utilize a mechanism corresponding to discrete movement generation. Huys et al. also analyzed the duration of each movement's acceleration and deceleration phase.

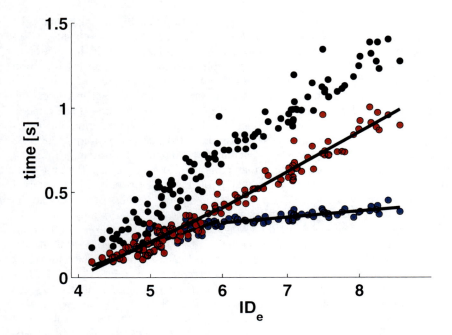

Fig. 3 Movement time, and acceleration and deceleration time as a function of ID. Movement time (black dots), acceleration time (blue dots) and deceleration time (red dots) as a function of ID for a single participant. The discontinuity in the acceleration time was identified by piece-wise linear regressions (represented by the black lines).

It turned out that for the rhythmic movements, the duration of both movement phases contributed approximately equally to the *MT* increase as a function of *ID*, while for the discrete movements the *MT* increase was almost exclusively due to an increase in the deceleration's phase duration, and the duration of the acceleration phase was close to constant as a function of *ID*. (The discontinuity in the acceleration phase was identified via regression analysis; see Huys et al., 2010 for details). The deceleration phase involved a discontinuity as a function of *ID*, which occurred at the rhythmic-discrete transition (see Fig. 3). By implication, Fitts' law involves a discontinuity. In a second experiment, it was shown that neither of these phenomena occurred when the participants performed similarly timed movements in the absence of the accuracy constraints; the movements were never generated in a discrete fashion. It appears, therefore, that humans abruptly engage a different control mechanism when task difficulty increases.

7 In Conclusion

The investigation of single limb movements from a dynamical perspective has been quite successful. As for the study of rhythmic movements, in terms of limit cycles, most predictions that can be derived form oscillator theory have been confirmed, even though a few detailed features that have shown up under a magnifying glass were not consistent with particular proposed nonlinear oscillators. Such little flaws were in some cases ingeniously 'repaired' in subsequent studies, as for instance the introduction of the frequency-dependent damping term by Beek et al., (1996) so as to improve the match of hybrid model to empirical data (see above). In other words, the conception of rhythmic movements in terms of limit cycle oscillators has proven to be justifiable and has, as a consequence, become widely adhered to. In addition, more recent approaches incorporating rhythmic and discrete movements have provided strong argument that both movements belong to classes that are not reducible to each other—or at least, it so appears. That is not to say that no criticism can be formulated — it has been, particularly so regarding the assumption that rhythmic movements can be fully described in terms of two-dimensional systems. In the early 1970s, Wing and Kristofferson (1973a, 1973b) developed a simple model that could produce a negative lag-one correlation between successive finger taps, indicating people's tendency to alternate the duration of successive taps around the mean tap interval. The negative (lag-one) correlation has turned out to be a very robust phenomenon, and has been reported for repetitive actions such as finger tapping (Hazeltine, Helmuth, & Ivry, 1997; Ivry, Keele, Diener, 1988; Wing & Kristofferson 1973a, 1973b), successive catches in juggling (Post, Daffertshofer, Beek, 2000; Huys, Daffertshofer & Beek, 2003) and saccadic eye movements (Collin, Jahanshahi, Barnes, 1998). Such repetitive actions are commonly conceived of as limit cycles, at least so by researchers in coordination dynamics. Daffertshofer (1998), however, analytically showed that two-dimensional autonomous limit cycles are unable to produce the negative lag-one correlation, irrespective of noise. At first blush, this suggests that repetitive movements revealing the negative lag-one correlation are thus controlled otherwise, be it by externally driven ('forced') limit cycles, by autonomous limit cycles of a dimension larger than two or as sequences of discrete movements. Experimental evidence, however, indicates that the negative lag-one correlation is present when the repetitive movements involve mechanical impact, as in tapping on a tabletop, but not in its absence, as in 'freely' oscillating one's finger (Delignières, Torre & Lemoine, 2008; Torre & Delignieres, 2008). The autonomous limit cycle description, however, breaks down in the presence of external mechanical impacts, as the latter provides a (spatial) discontinuity in the phase flow and haptic input. In other words, Daffertshofer's result is in line with experimental findings but does not by itself imply that an autonomous two-dimensional limit cycle description for rhythmic movements is insufficient. Nevertheless, the assumption of (complete) autonomy — at least so at the level of limb motion — has also been questioned by (among others) Kay and colleagues (1991) and Beek (cf. Beek, 1989; Beek & Beek, 1988; see also Kay, 1988, for a dimensionality analysis of rhythmic movements). Thus, while the two-dimensional description has proven to be a good approximation, which, in conjunction with the manageability of two-dimensional relative to higher dimensional systems, undoubtedly explains the widespread tendency to stick to them, it cannot account for all experimentally observed phenomena pertaining to single rhythmic limb movements. In addition, for discrete movements an autonomous two-dimensional description is out of the question. Should this be taken to imply that the dynamical solution to Bernstein's degree-of-freedom problem has failed? Although it would be premature to formulate a definitive answer, our gist is "no". For those instances in which full autonomy cannot be established in two dimensions — and it may well be that such will be the case for many instances under a scrutinizing view — the challenge is to identify at what level of description additional dimensions should be incorporated. One example of an attempt thereto was briefly discussed above for the case of juggling (Beek, 1989); others can be found in Beek et al., (2002) and Jirsa et al. (1998). In other words, while some important issues remain to be resolved, it seems fair to say that the dynamical approach to motor control and coordination has proven its worth and may be trusted to continue to deliver novel insights for years to come.

References

Adam, J.J., Paas, F.: Dwell time in reciprocal aiming tasks. Human Movement Science 15, 1–24 (1996)
Beek, P.J.: Juggling dynamics. Free University Press, Amsterdam (1989)
Beek, P.J., Beek, W.J.: Tools for constructing dynamical models of rhythmic movement. Human Movement Science 7, 301–342 (1988)
Beek, P.J., Rikkert, W.E.I., van Wieringen, P.C.W.: Limit cycle properties of rhythmic forearm movements. Journal of Experimental Psychology: Human Perception and Performance 22, 1077–1093 (1996)
Beek, P.J., Schmidt, R.C., Morris, A.W., Sim, M.Y., Turvey, M.T.: Linear and nonlinear stiffness and friction in biological rhythmic movements. Biological Cybernetics 73, 499–507 (1995)
Beek, P.J., Peper, C.E., Daffertshofer, A.: Modeling rhythmic interlimb coordination: Beyond the Haken-Kelso-Bunz model. Brain & Cognition 48, 149–165 (2002)
Bernstein, N.: The coordination and regulation of movement. Pergamon Press, London (1967)
Bootsma, R.J., Boulard, M., Fernandez, L., Mottet, D.: Informational constraints in human precision aiming. Neuroscience Letters 333, 141–145 (2003)
Buchanan, J.J., Park, J.-H., Ryu, Y.U., Shea, C.H.: Discrete and cyclical units of action in a mixed target-pairing task. Experimental Brain Research 150, 473–489 (2003)
Calvin, S., Jirsa, V.K.: Perspectives in the dynamic nature of coupling in human coordination. In: Huys, R., Jirsa, V.K. (eds.) Nonlinear Dynamics in Human Behavior. SCI, vol. 328, pp. 91–114. Springer, Heidelberg (2010)
Collins, C.J., Jahanshahi, M., Barnes, G.R.: Timing variability of repetitive saccadic eye movements. Experimental Brain Research 120, 325–334 (1998)
Daffertshofer, A.: Effects of noise on the phase dynamics of nonlinear oscillators. Physical Review E 58, 327–338 (1998)
Daffertshofer, A.: Benefits and pitfalls in analyzing noise in dynamical systems – on stochastic differential equations and system identification. In: Huys, R., Jirsa, V.K. (eds.) Nonlinear Dynamics in Human Behavior. SCI, vol. 328, pp. 35–68. Springer, Heidelberg (2010)

Delignières, D., Torre, K., Lemoine, L.: Fractal models for event-based and dynamical timers. Acta Psychologica. 127, 382–397 (2008)

Dennett, C.D.: Brainstorms: Philosophical Essays on Mind and Psychology. Bradford Books (1978)

Dennett, C.D.: Consciousness Explained. The Penguin Press, Harmondsworth (1991)

Elliott, D., Helsen, W., Chua, R.: A Century Later: Woodworth's (1899) Two-Component Model of Goal-Directed Aiming. Psychological Bulletin 127, 342–357 (2001)

Fink, P.W., Kelso, J.A.S., Jirsa, V.K.: Perturbation-induced false starts as a test of the jirsa-kelso excitator model. Journal of Motor Behavior 41, 147–157 (2009)

Fitts, P.M.: The information capacity of the human motor system in controlling the amplitude of movement. Journal of Experimental Psychology 47, 381–391 (1954)

Fitts, P.M., Peterson, J.R.: Information capacity of discrete motor responses. Journal of Experimental Psychology 67, 103–112 (1964)

FitzHugh, R.: Impulses and physiological states in theoretical models of nerve membrane. Biophysical Journal 1, 445–466 (1961)

Fuchs, A.: Dynamical systems in one and two dimension: a geometrical approach. In: Huys, R., Jirsa, V.K. (eds.) Nonlinear Dynamics in Human Behavior. SCI, vol. 328, pp. 1–33. Springer, Heidelberg (2010)

Gonzalez, D.L., Piro, O.: Global bifurcations and phase portrait of an analytically solvable nonlinear oscillator: relaxation oscillations and saddle-node collisions. Physical Review A 36, 4402–4410 (1987)

Guckenheimer, J., Holmes, P.: Nonlinear Oscillations, Dynamical Systems, and Bifurcations of Vector Fields. Springer, New York (1983)

Guiard, Y.: On Fitts's and Hooke's laws: Simple harmonic movement in upper-limb cyclical aiming. Acta Psychologica 82, 139–159 (1993)

Guiard, Y.: Fitts' law in the discrete vs. continuous paradigm. Human Movement Science 16, 97–131 (1997)

Goodwin, B.C.: Temporal Organisation in cells. Academic Press, New York (1963)

Goodwin, B.C.: Biological stability. In: Waddington, C.H. (ed.) Towards a Theoretical Biology (1970)

Haken, H.: Synergetics. An introduction: nonequilibrium phase transitions and self-organization in physics, chemistry, and biology. Springer, Berlin (1977)

Haken, H.: Synergetics: an introduction: nonequilibrium phase transitions and self-organization in physics, chemistry, and biology (3rd rev. and enl. ed.). Springer, Berlin (1983)

Haken, H.: Principles of brain functioning. A synergetic approach to brain activity, behavior and cognition. Springer, Heidelberg (1996)

Haken, H., Kelso, J.A.S., Bunz, H.: A theoretical model of phase transitions in human hand movements. Biological Cybernetics 51, 347–356 (1985)

Hazeltine, E., Helmuth, L.L., Ivry, R.B.: Neural mechanisms of timing. Trends in Cognitive Science 1, 163–169 (1997)

Hogan, N., Sternad, D.: On rhythmic and discrete movement: Reflections, definitions and implications for motor control. Experimental Brain Research 181, 13–30 (2007)

Huys, R., Daffertshofer, A., Beek, P.J.: Learning to juggle: on the assembly of functional subsystems into a task-specific dynamical organization. Biological Cybernetics 88, 302–318 (2003)

Huys, R., Studenka, B.E., Rheame, N., Zelaznik, H.N., Jirsa, V.K.: Distinct timing mechanisms produce discrete and continuous movements. PLoS Computational Biology 4, e1000061 (2008)

Huys, R., Fernandez, L., Bootsma, R.J., Jirsa, V.K.: Fitts' Law is not continuous in reciprocal aiming. Proceedings of the Royal Society B: Biological Sciences 277, 1179–1184 (2010)

Iberall, A.S.: Periodic phenomena in organisms seen as non-linear systems. Theoria to Theory 4, 40–53 (1970)

Iberall, A.S.: A field and circuit thermodynamics for integrative physiology: I. Introduction to general notion. American Journal of Physiology: Regulatory, Integrative and Comparative Physiology 2, R171– R180 (1977)

Iberall, A.S.: Power and communicational spectroscopy in biology. American Journal of Physiology: Regulatory, Integrative and Comparative Physiology 3, R3–R19 (1978)

Ivry, R.B., Keelse, S.X., Diener, H.C.: Dissociation of the lateral and medial cerebellum in movement timing and movement execution. Experimental Brain Research 73, 167–180 (1988)

Jirsa, V.K., Fuchs, A., Kelso, J.A.S.: Connecting cortical and behavioral dynamics: bimanual coordination. Neural Computation 10, 2019–2045 (1998)

Jirsa, V.K., Kelso, J.A.S.: The excitator as a minimal model for the coordination dynamics of discrete and rhythmic movement generation. Journal of Motor Behavior 37, 35–51 (2005)

Jordan, D.W., Smith, P.: Nonlinear Ordinary Differential Equations: An Introduction to Dynamical Systems. Oxford University Press, Oxford (1999)

Kay, B.A.: The dimensionality of movement trajectories and the degrees of freedom problem: A Tutorial. Human Movement Science 7, 343–364 (1988)

Kay, B.A., Kelso, J.A.S., Saltzman, E.L., Schöner, G.: Space-time behavior of single and bimanual rhythmical movements: data and limit cycle model. Journal of Experimental Psychology: Human Perception & Performance 13, 178–192 (1987)

Kay, B.A., Saltzman, E.L., Kelso, J.A.S.: Steady-state and perturbed rhythmical movements: a dynamical analysis. Journal of Experimental Psychology: Human Perception & Performance 17, 183–197 (1991)

Kelso, J.A.S.: Dynamic Patterns: the Self-Organisation of Brain and Behavior. MIT Press, Cambridge (1995)

Kelso, J.A.S., Holt, K.G., Kugler, P.N., Turvey, M.T.: On the concept of coordinative structures as dissipative structures: II. Empirical lines of convergence. In: Stelmach, G.E., Requin, J. (eds.) Tutorials in motor behavior, pp. 49–70. North-Holland, Amsterdam (1980)

Kelso, J.A.S., Holt, K.G., Rubin, P., Kugler, P.N.: Patterns of human interlimb coordination emerge from the popserties of non-linear, limit cycle oscillatory processes: Theory and data. Journal of Motor Behaviour 13, 226–261 (1981)

Kugler, P.N., Kelso, J.A.S., Turvey, M.T.: On the concept of coordinative structures as dissipative structures: I. Theoretical lines of convergence. In: Stelmach, G.E., Requin, J. (eds.) Tutorials in Motor Behavior. North-Holland, Amsterdam (1980)

Murray, J.D.: Mathematical Biology. Springer, New York (1993)

Meijer, O.G., Roth, K.: Complex Movement Behavior: The motor-action controversy. Elsevier Science Publishers BV, North-Holland (1988)

Meyer, D.E., Abrams, R.A., Kornblum, S., Wright, C.E., Smith, J.K.: Optimality in human motor performance: Ideal control of rapid aimed movements. Psychological Review 95, 340–370 (1988)

Mottet, D., Bootsma, R.J.: The dynamics of goal-directed rhythmical aiming. Biological Cybernetics 80, 235–245 (1999)

Mottet, D., Bootsma, R.J.: The dynamics of rhythmical aiming in 2D task space: Relations between geometry and kinematics under examination. Human Movement Science 20, 213–241 (2001)

Nicolis, G., Prigogine, I.: Exploring complexity: An introduction. W H Freeman and Co., New York (1989)

Pattee, H.H.: The hierarchical nature of controls in living matter. In: Rosen, R. (ed.) Foundations of mathematical biology, vol. 1, pp. 1–22. Academic Press, New York (1972)

Pattee, H.H.: The physical basis and origin of hierarchical control. In: Pattee, H.H. (ed.) Hierarchy theory: The challenge of complex systems, pp. 71–108. George Braziller, Inc., New York (1973)

Perko, L.: Differential Equations and Dynamical Systems. Springer, New York (1991)

Prigogine, I.: Structure, dissipation and life. In: Marois, M. (ed.) Theoretical physics and biology, pp. 23–52. North-Holland, Amsterdam (1969)

Prigogine, I., Nicolis, G.: Self-Organization in Non-Equilibrium Systems. Wiley, Chichester (1977)

Post, A.A., Daffertshofer, A., Beek, P.J.: Principal components in three-ball cascade juggling. Biological Cybernetics 82, 143–152 (2000)

Rosenbaum, D.A.: Is dynamical systems modelling just curve fitting? Motor Control 2, 101–104 (1998)

Schmidt, R.A.: A schema theory of discrete motor skill learning. Psychological Review 82, 225–260 (1975)

Schmidt, R.A., Lee, T.D.: Motor Control and Learning: A Behavioral Emphasis. Human Kinetics, Urbana (2005)

Schöner, G.: A dynamic theory of coordination of discrete movement. Biological Cybernetics 53, 257–270 (1990)

Schöner, G., Kelso, J.A.S.: A synergetic theory of environmentally-specified and learned patterns of movement coordination. I. Relative phase dynamics. Biological Cybernetics 58, 71–80 (1988)

Sternad, D.: Towards a unified theory of rhythmic and discrete movements–Behavioral, modeling and imaging results. In: Fuchs, A., Jirsa, V.K. (eds.) Coordination: Neural, Behavioral and Social Dynamics, pp. 105–133 (2007)

Steward, I.: Concepts of modern mathematics. Dover Publications, New York (1995)

Strogatz, S.H.: Nonlinear dynamics and chaos. With applications to physics, biology, chemistry, and engineering. Cambridge Massachusetts, Perseus (1994)

Torre, K., Delignières, D.: Distinct ways for timing movements in bimanual coordination tasks: The contribution of serial correlation analysis and implications for modelling. Acta Psychologica 129, 284–296 (2008)

Waddington, C.H. (ed.): Towards a Theoretical Biology 1: Prolegomena. Aldine, Birmingham (1968)

Waddington, C.H. (ed.): Towards a Theoretical Biology 2: Sketches. Edinburgh Univ. Press, Edinburgh (1969)

Waddington, C.H. (ed.): Towards a Theoretical Biology 3: Drafts. Edinburgh Univ. Press, Edinburgh (1970)

Waddington, C.H. (ed.): Towards a Theoretical Biology 4: Essays. Edinburgh Univ. Press, Edinburgh (1972)
Winfree, A.T.: The Geometry of Biological Time. Springer, New York (1980)
Wing, A.M., Kristofferson, A.B.: The timing of interresponse intervals. Perception & Psychophysics 13, 455–460 (1973a)
Wing, A.M., Kristofferson, A.B.: Response delays and the timing of discrete motor responses. Perception & Psychophysics 13, 5–12 (1973b)
Woodworth, R.S.: The accuracy of voluntary movements. Psychological Review 3, 1–106 (1899)

Perspectives on the Dynamic Nature of Coupling in Human Coordination

Sarah Calvin and Viktor K. Jirsa

Abstract. This chapter focuses on motor coordination between similar as well as different classes of movements from a phase flow perspective. Most studies on coordination dynamics are concerned with coordination of rhythmic movements. This constraint enables the modeler to describe the interaction between the oscillating movement components by a phase description, and its dynamics by a potential function. However, potential functions are extremely limited and exist only in a limit number of cases. In contrast, dynamical systems can be unambiguously described through their phase flows. The present chapter elaborates on coordination dynamics from the phase flow perspective and sheds new light on the meaning of biological coupling. The phase flow deformations of coupled systems may be understood using the notion of convergence and divergence of the phase space trajectories and aid in explaining the mechanisms of trajectory formation and the interaction (coupling) between arbitrary movements.

On the Nature of Coordination

1 Introduction

The human body can be understood as a complex system, since it is composed of millions of cells organized in tissues that form organs responsible for basic behavior such as breathing and walking, but also cognitive and complex behavior such as speaking, reasoning, thinking, learning and so on. A multitude of components interact together at any level (molecular, cellular, tissular, behavioral, psychological, social ...). An important question is how such a high-dimensional system can be organized, at least temporarily, to achieve a given (task) goal. For Bernstein (1967), it is a problem of reducing the number of variables to be controlled in

---

Sarah Calvin · Viktor K. Jirsa
Theoretical Neuroscience Group
Université de la Méditerranée, UMR 6233 "Movement Science Institute",
CNRS, Faculté des Sciences du Sport, 13288, Marseille cedex 09, France

mastering the several degrees of freedom involved. He hypothesized that they might be — temporarily or not — (dis)integrated into a functional coordinative structure (unit). Then, functionally speaking, the coordination is probably the most efficient tool that nature disposes to manage the complexity. Indeed, coordination implies that two or more subsystems work together rather than independently: we call this coupling. From that cooperation between the different parts, patterns of behavior emerge. How can the coordinated behavior of complex biological systems to be understood? Among others, the dynamical pattern theory using concepts and methods gleaned from sciences of complexity (Synergetics; Haken (1977)) and theories of nonequilibrium phase transitions, proposed a theoretical framework to describe laws and principles that govern pattern formation and coordination. The emergence of patterns is possible only in nonlinear[1] complex systems that are open (i.e., interacting with environment), far from equilibrium known to exhibit some properties such as self-organization. This property allows these systems to exhibit spontaneous transition between different states and patterns of activity, as a result of internal regulation in response to changes in external condition that are unspecific to the emerging pattern Haken (1983). These states or patterns are "attractors"[2] and correspond to a stable mode of behavior that the system tends to spontaneously adopt and maintain depending on the constraints applied on it.

Despite the complexity of a system's behavior, it is still often possible to provide a low-dimensional description of its evolution. To do so, it is necessary to isolate a set of variables, the so-called collective variables or order parameters. These variables completely characterize the emerging patterns in the system, and hence reduce the degrees of freedom available to the system (Haken 1983; Kelso 1995). The identification of collective variables is a longstanding issue and typically cannot be addressed a priori by the experimenter; they are often predominant near regions of pattern change i.e., when the stability of an ongoing coordination pattern changes and instabilities occur. These instabilities are induced by change of a so-called control parameter that leads the system in an unspecific manner through a series of different states and state transitions. Indeed, when the control parameter crosses a critical value, instabilities occur leading to the emergence of new coordinative patterns different from the formerly ones, and to the disappearance of previous ones. Once identified, the examination of the dynamics of the collective variable (i.e., how does the collective variable change as the control

---

[1] The nonlinearity can be understood as a loss of the proportionality of cause and consequence. Thus differently to linear systems, the behavior of a nonlinear system cannot be inferred and even constructed from the behavior of its components ("*the whole is more than and different from the sum of the parts*", Phil Anderson (1972)). The nonlinearity provides the complex system with the possibility to exhibit different qualitative and multiple behaviors.

[2] An attractor is a structure toward which all the trajectories in state space converge as time goes to infinity. Different kinds of attractors are known: a/ (stable or unstable) fixed point, b/ (stable or unstable) limit cycle when the system exhibits a periodic behavior and c/ strange attractors when the system displays a chaotic, unpredictable quasi-cyclical behavior. The state space is defined by the set of the system's state variables.

parameter varies) allows to understand how (and which) coordinated behaviors are produced in complex systems. For example, it is possible to examine if the order parameter tends to adopt specific value(s) and, in other words, if the system converges to one (or several) attractor(s) as time goes to infinity, but also if some values of the order parameter act as repellor(s), i.e., an unstable state from which the system evolves away. From that, dynamical laws can be written; they usually correspond to a set of mathematical (differential) equations formalizing the temporal evolution of the collective variable(s). These laws are helpful for studying, understanding and predicting how a system behaves. A common way to represent attractors and repellors is to symbolize them as "valleys" and "hills", respectively (see figure 1) by using a potential. To model the intrinsic dynamic of a system, we can imagine a ball in a landscape with valleys and hills. Once the ball leaves its initial condition on top of a hill, it will slide down into the "valley" and away from the top of the "hill", i.e., the repellor. The part of the landscape between a repellor and an attractor is the basin of attraction. The deeper the valley, the stronger (and consequently the more stable) is the attractor.

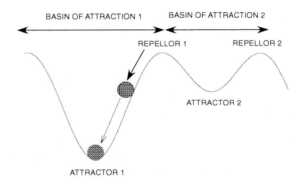

Fig. 1 Representation of a dynamics with attractors and repellors using a potential. The attractors correspond to the valleys, and the repellors to the hill tops of the landscape. The ball symbolizes the evolution of the system toward the attractor as time evolves.

In the movement domain, motor behavior such as motor coordination and more specifically rhythmic coordination is viewed as self-organized pattern formation processes and studied from the dynamics perspective Kelso (1995). Scott Kelso named this line of thinking Coordination Dynamics, which finds its roots and mathematical formalization in the more general field of synergetics pioneered by Hermann Haken (1977, 1983). In this framework, coordination is viewed as an emergent product from the interaction between components. In the human, experimental observations showed that coordinated movements are spontaneously performed following characteristic stable patterns of behavior. These stable patterns (or modes), however, exist in small numbers. For instance, when asked to move rhythmically two hands or fingers at the same pace, humans produce only two patterns with relative phasing, ϕ, of both limbs equal to 0° and 180° referred to as in-phase and anti-phase patterns, respectively (Tuller et al., 1989; Yamanishi et al., 1979). These observations revealed that without any learning or guidance, it is almost impossible to naturally produce rhythmic coordination patterns with different relative phasing other than 0° and 180° as if values outside of these distinct (stable) states were inaccessible Turvey (1990). Consequently, questions arise why complex systems composed of muscles, tendons, joints and neurons can produce spontaneously only two stables modes among multiple, in fact infinitely many, possibilities. The mean principles underlying the emergence of coordination together with their modeling will be treated in the next paragraph.

2 Coordination Depends on Intrinsic Dynamics and Coupling

To understand how coordination comes about in two coupled systems (indexed by 1 and 2 in the following), it is useful to provide a general formulation:

$$\begin{cases} \dot{x}_1 = y_1 \\ \dot{y}_1 = -x_1\omega_1^2 + f(x_1, y_1) + C(x_1, y_1, x_2, y_2) \end{cases}$$

$$\begin{cases} \dot{x}_2 = y_2 \\ \dot{y}_2 = \underbrace{-x_2\omega_2^2 + f(x_2, y_2)}_{\text{intrinsic dynamics}} + \underbrace{C(x_2, y_2, x_1, y_1)}_{\text{coupling}} \end{cases}$$

The four state variables, positions x_1, x_2 and velocities y_1, y_2, define a four-dimensional system and hence the system "lives" in a four-dimensional state space. The dots on the variables denote the time derivative and ω_1, ω_2 are the preferred frequencies (eigenfrequencies). In conjunction, the eigenfrequencies and the function f define the intrinsic dynamics. The coupling between components is given by the function C. Evidently, the system behavior depends (only) on the intrinsic dynamics and the coupling amongst the components.

Following Haken, Kelso, Bunz (1985), let us imagine that the coupled systems are nonlinear hybrid oscillators[3] and that the coupling is nonlinear. Within a state space, the coordinates of which correspond to the oscillators' state variables, i.e., position and velocity, the dynamics of each oscillator takes the form of a circle (limit cycle), which is being traced out periodically (see figure 2). The angular position on the circle is called the phase. Since movements in mechanics are often periodic, the phase plays an important role and the state space is often referred to as the phase space. When uncoupled, each oscillator can take any phase value or, equivalently, any relative phase, $\phi = \phi_1 - \phi_2$, between the two oscillators is possible.

---

[3] In motor control, much of our knowledge on laws and principles governing coordination dynamics originates from studies involving rhythmic movements. Movements of individual limb involved in a periodical activity are used to be modeled as a two-dimensional self-sustained hybrid oscillator. This kind of oscillator exhibits a decrease in amplitude together with an increase of peak velocity when the frequency is increased. The term hybrid comes from the fact that rhythmic biological movements are modeled as oscillators characteristics of which are a combination of the well-known Rayleigh and van der Pol oscillators.

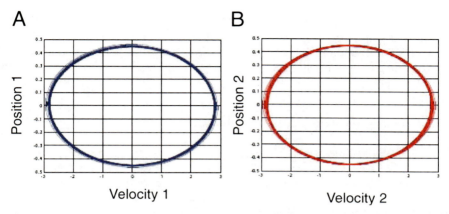

Fig. 2 Phase portrait of two uncoupled hybrid oscillators. The trajectories of each oscillator lie on the limit cycle (A, B) during rhythmic movements.

Once the two hybrid oscillators are coupled, coordination emerges. From this coupling, two phenomena arise (in its simplest case). First, it induces a frequency locking of the two oscillators; they start to oscillate with a common frequency, intermediate to their own preferred frequencies. Second, it limits the relative phase values that the two oscillators can take: for the in-phase relationship, the coupling brings the phases of the two oscillators together, whereas for the anti-phase relationship, the two phases are in opposition. In other words, the coupling induces (or mediates) phase-attraction and phase-repulsion processes. Once the coordination is established, its dynamics (i.e., the number and the type of emerging patterns on one hand, and how these patterns evolve in time, on the other hand) will depend not only on the nature of the coupling but also on the intrinsic characteristics of each involved component.

2.1 Effect of the Coupling

Multistability and multifunctionality of a complex system necessitate nonlinearity, which may be captured either in the intrinsic dynamics of the system or its coupling function. In the Haken–Kelso–Bunz (HKB) system, both contributions are nonlinear; in particular, the coupling function C reads

$$C(x_2, y_2, x_1, y_1) = (\dot{x}_{12} - \dot{x}_{2,1})(\alpha + \beta(x_{1,2} - x_{2,1})^2)$$
$$= \underbrace{\alpha(\dot{x}_{12} - \dot{x}_{2,1})}_{A} + \underbrace{(\beta(x_{1,2} - x_{2,1})^2(\dot{x}_{12} - \dot{x}_{2,1}))}_{B}$$

In this equation, the term A creates the presence of the attractor ϕ=0° and the term B contributes to both, the creation of the attractor at ϕ=0° and 180°. When the frequency increases, the amplitude of the trajectories of the two oscillators decreases and results in a decrease of $\beta(x_{1,2}-x_{2,1})^2$ if $\phi=180°$, thus reducing the term B and consequently reducing the stability of the 180° attractor. The $\phi=0°$ attractor is unaffected. At the critical value of frequency, the mode of coordination corresponding to the anti-phase pattern no longer exits, and a transition to in-phase occurs.

2.2 Effect of the Intrinsic Dynamics

The second critical element is the intrinsic dynamics of the coordinated element itself, that is its inherent characteristic behavior in the absence of any coupling. It is easy to understand when considering that an interaction between elements underlies the confrontation of two following mechanisms: The first is the tendency, called M(agnet) effect Von Holst (1973), each oscillator displays to attract the other to its own frequency, and the second is opposed and constitutes the maintenance tendency in which each component tends to stay at its personal tempo.

2.2.1 Coupling Components with Identical Intrinsic Dynamics

Let us imagine the two coupled oscillators, each oscillating following an eigenfrequency ω_1 and ω_2, respectively. The detuning, that is the difference between the two oscillators, can be calculated as follows $\Delta\omega=\omega_2-\omega_1$. Of course, when $\Delta\omega=0$, the components do not differ and oscillate individually following the same cycling rate $\omega=\omega_1=\omega_2$. In this case, the M effect is strong and the coordination is frequency- and phase-locked. The former means oscillation with the same frequency, and the latter indicates that the phase difference of the oscillators is fixed. In the following, we discuss coupled systems with identical intrinsic dynamics.

A classic example is provided by the coordination of two index fingers (or any other limbs) by means of a biological coupling. The dynamics of this kind of coordination is now well documented in the field of the motor control domain since the works of Kelso (1981, 1984), who first investigated the idea that motor coordination may be characterized by principles of self-organization, giving birth to the dynamical pattern perspective[4]. These experiments examined bimanual coordination between finger oscillatory movements and created the paradigm of the "wriggling fingers". Moreover, they also allowed the elaboration of the first model, the well-known HKB model Haken (1985). Subjects were asked to oscillate their index fingers with the same frequency. With no training whatsoever, they were able to comfortably perform two coordination patterns. The in-phase pattern was produced by the co-activation of the homologous muscles, so that the two fingers flexed and extended together led to a phasing of 0° between the limbs. The anti-phase was

---

[4] The dynamical pattern perspective is a theoretical framework used for more than 20 years. It aims to identify laws and principle underlying pattern formation in the motor domain and is grounded in the theory of complex systems and synergetics.

produced by the co-activation of the non-homologous muscles so that one finger was flexed as the other one was extended leading to a phasing of 180°. The dynamics of these preferred coordination modes was explored by increasing the frequency of oscillation. When the coordination was first initiated in-phase, a relative phase was maintained around 0° in a stable fashion (i.e., with a minimal variability) when the frequency increased. On the other hand, when the subjects started to oscillate in anti-phase, the increase of the imposed frequency made a 180° phasing between the fingers difficult to maintain, so that abrupt switching to the in-phase patterns occurred when the frequency crossed critical values. Before this phase transition, the variability of the relative phase (required at 180°) between the two fingers was dramatically enhanced as the frequency increased. However, after the transition from the anti-phase to the in-phase, the variability of the relative phase was low and comparable to that produced during a pattern directly prepared in-phase. This phenomenon is referred to as critical fluctuations and is indeed an indicator of being near a threshold of a phase transition. The observation of critical fluctuations is one of the strongest arguments supporting the interpretation of the "wiggling finger" transition as a phase transition. Other explanations would have difficulty in explaining naturally their occurrence.

In summary, these experiments provided several key insights. First, the relative phase between coordinated segments qualifies as a good candidate for the status as a collective variable since it expresses the relationships between the different parts (here, the two fingers) and reports the dynamics of these patterns and particularly their changes. Second, the frequency serves as a control parameter in this paradigm. In other words, the control parameter controls the stability of the different coordinative patterns. Third, the existence of critical fluctuations underwrites the nature of the transition mechanism to be a bifurcation and a generic mechanism of pattern switching.

Inspired by these experimental findings, Haken, Kelso and Bunz (1985) proposed a model not only explaining Kelso's experimental results, but also making new predictions. This model captured with a one-dimensional dynamical equation the phenomena experimentally observed by Kelso (1981, 1984). More precisely, the model formalized the rhythmic interlimb coordination at two interconnected levels: a system of two (identical) coupled nonlinear oscillators describing the individual limb movements and, consistent herewith, a so-called potential function that describes the dynamics of the collective variable interpreted as the relative phase (ϕ) between the oscillators Haken (1985). The potential function and its dependence on the control parameter are shown in figure 3.

Remember, the experimental findings reported that at a low frequency regime, ϕ can take two stable values: $\phi=0°$ and $\phi=180°$. These two stationary states — in other words the bistability of the system — were modeled as two fixed point attractors. To account for the transition, the bistability must change toward monostability (where only $\phi=0°$ is stable) when the frequency reaches a critical value. Since symmetry exists between left and right limb movements, the model must be able to describe the same coordination dynamics under the exchange of indices.

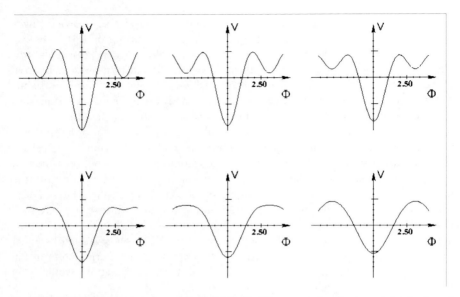

Fig. 3 The HKB model. A series of potential fields (from top left to bottom right) is plotted for different values of the control parameter. The upper left potential describing the coordination dynamics at the slowest frequency: the attractors corresponding to Φ=0° and Φ =180° (here, equivalent to its value in radians, π) are present as minima in the potential. As the frequency increases (from left to right, upper row first), the minimum at Φ =180° becomes more shallow and finally disappears, leaving a maximum at its place.

To model the dynamics of the order parameter, Haken et al. proposed the dynamics

$$\dot{\phi} = -\frac{dV(\phi)}{d\phi}$$

where $\dot{\phi} = -a\sin(\phi) - 2b\sin 2\phi$ and the potential function $V(\phi) = -a\cos(\phi) - b\cos(2\phi)$, and a and b are constants. With this potential function formulation, the temporal evaluation of ϕ corresponds to the damped motion of a mass in the potential landscape, with valleys and peaks representing stable and unstable solutions, respectively Haken (1985). The b/a ratio simulates the change in frequency (i.e., the increase of the control parameter). Changing the ratio from 1 to 0 corresponds to increase in the frequency rate. As shown in figure 5, modifying this ratio induces a change in the shape of the potential reflecting the dynamics of ϕ. More precisely, when the ratio is close to 1, the dynamics is bistable and the potential exhibits two minima (attractors) centered on 0° and ± 180°. The minimum corresponding to 0° is deeper and more straight than that corresponding to 180°, reflecting the differential stability of the in-phase and anti-phase patterns. When b/a goes to 0, the 180° valley progressively vanishes and disappears totally for high-frequency cycling. Then the system (here, two coupled fingers) cannot any longer stay in this 180° valley and is attracted to the 0° valley (still present, but shallower). This reflects the anti-phase to in-phase pattern phase transition. Imagine now that the system is first located at the bottom of the 0° valley, meaning that the system begins to oscillate in-phase. Because of the depth of the well, the system cannot jump out of the 0° basin of attraction. Later stochastic forces were added to account for the mechanisms responsible for the phase transition, which is the critical fluctuations of the order parameter that are elicited by the interaction of the different parts of the system Schöner (1990).

2.2.2 Coupling Components with Different Intrinsic Dynamics

2.2.2.1 Similar Oscillators with Different Eigenfrequency
How does the coordination dynamics evolve when the coupled elements do not share the same characteristics? In terms of behavior, dynamical phenomena (pattern formation, multistability, phase transition, critical fluctuations ...), first identified by Kelso (1981, 1984), turned out to be generic and were reported when coupling oscillators of various nature, i.e., when coupling a limb and an auditory metronome (Kelso et al., 1990; Wimmers et al., 1992) or in case of coordination between two (or more) non-homologous (a wrist and an ankle, for instance) limbs (Kelso et al., 1992; Kelso et al., 1991; Salesse et al., 2005) with different length, size and weight, and consequently.

In such coordination, components exhibit different individual eigenfrequencies, $\Delta\omega$ differs from zero. More precisely, the higher $\Delta\omega$, the more heterogenic are the components. This is of importance because a stable synchronization (that is constantly maintaining a phase- and frequency-locked coordination) can only be produced for a certain range of $\Delta\omega$ values (see figure 6). For concreteness, let us imagine a one-dimensional system $\dot{\phi} = \Delta\omega - c\sin\phi$ with $\phi = \phi_1 - \phi_2$, $\Delta\omega = \omega_1 - \omega_2$ and c the coupling function. To find the phase-locked states, we set $\dot{\phi} = 0 \Leftrightarrow \Delta\omega - c\sin\phi = 0 \Rightarrow |\sin\phi| = \left|\frac{\Delta\omega}{c}\right| < 1$. In a more general manner, figure 4 clearly illustrates that the synchronization and phase-locking phenomena depend on both the coupling (c) and the intrinsic dynamics of each oscillator ($\Delta\omega$).

Even if the M effect and the attraction to a certain phase still exist, they are strongly counterbalanced, even dominated by the maintenance tendency. Then in this case, ϕ can take a large range of values giving birth to a new kind of behavior labeled "relative" coordination. The latter is characterized by the fact that it does not correspond to a mode locking. Then, due to the detuning, we can observe systematic and progressive slippage in the phase relation between components interrupted by stationary stages of phase locking. The latter has been termed intermittency by Kelso. The potential function $V(\Phi) = -\Delta\omega - a\cos(\Phi) - b\cos(2\Phi)$ for various values of detuning, $\Delta\omega$, is plotted in figure 5.

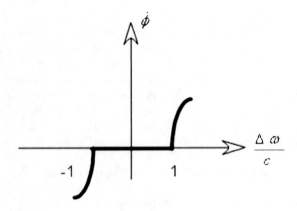

Fig. 4 The rate of change of Φ is plotted against $\frac{\Delta\omega}{c}$. The figure shows that the oscillators are phase locked ($\dot{\phi}=0$) only when $\frac{\Delta\omega}{c}$ varies between -1 and 1.

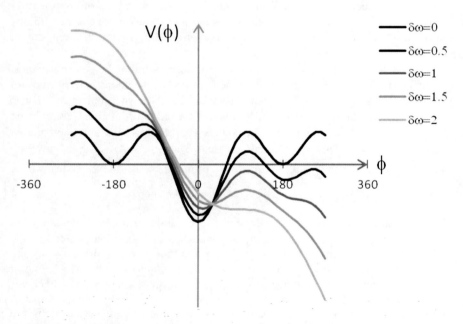

Fig. 5 Representation of the potential V(f) with broken symmetry (Δω). The amount of detuning (Δω) together with the frequency increase (from black to light gray lines, b/a=1, 0.75, 0.5, 0.25, 0) induces the progressive vanishing of stable fixed points located at ϕ=0, ±180° together with a shifting away from the pure in-phase (0°) and anti-phase patterns (180°).

2.2.2.2 Interaction of Movement of Different Type

So far, we have focused on coordination involving oscillatory movements. However, simultaneously moving two limbs in a rhythmical fashion constitutes only a subset of human motor behavior, especially so when considering the upper limbs. Most of the time our limbs move either in a discrete or rhythmic fashion, or in a combination thereof. Discrete movements are segmented motions preceded and followed by a period of quiescence where the velocity is zero. Such endpoints do not exist when executing periodical movements and rather appear smooth and regular, often sinusoidal. Using a classification based on phase flow topologies, recent works of Huys and colleagues Huys (2008) unambiguously showed that discrete and rhythmic movements pertain to different classes. The discrete movement can be described in terms of fixed point dynamics, whereas the rhythmic movement can be described in terms of limit cycle. These two topologically different structures cannot be reduced to each other and have consequently a different (intrinsic) dynamics[5] (Perko, 1991; Strogatz, 1994 (see also Raoul Huys in this volume)]. Various studies, in particular led by Dagmar Sternad and colleagues, have investigated the interaction between simultaneously executed arbitrary movements such as discrete and rhythmic movements (Latash, 2000; Michaels et al., 1994; Sternad, 2008; Sternad et al., 2007; Sternad et al., 2000; Sternad et al., 2007; Wei, 2003). The results of these studies are many, however, not univocal, because often appropriate control conditions were lacking or were different across different studies; frequently only limited ranges of frequency were examined, and it has not been clear if the adopted variables characterized the system dynamics sufficiently. In other words, the coordination dynamics of non-rhythmic movement is far less understood than that of rhythmic movements. This is not surprising considering that rhythmic movements are sufficiently described by its phase variable, which is not trivially the case for discrete movements. In other words, rhythmic movements appear to be lower dimensional than discrete movements. In a recent study, some of us Calvin (2010) have examined the coupling between discrete and rhythmic movements through an behavioral experiment. We manipulated the position in the rhythmic cycle of 'discrete stimulus deliverance' as well as the rhythmic movement frequency. Participants performed auditory-paced left-handed oscillations around the wrist at 0.5 Hz, 1.0 Hz, 1.5 Hz, 2.0 Hz and 2.5 Hz. They were instructed to perform a single (i.e., discrete) flexion–extension movement as fast as possible at the onset of a auditory stimulus that was given at different phases relative to the left-handed oscillations (at 0°, 45°, 90°, 135°, 180°, 225°, 270° and 315°). Analyses indicated that the coupling affects both the discrete and the rhythmic movement. Indeed, reaction time (RT) and movement time (MT) of the discrete hand movement was longer as the frequency of oscillation of the contralateral left hand was slower. No effect of the phase of discrete movement initiation was found. In contrast, the (normalized) Hilbert phase progression (indicating the degree of acceleration/deceleration) of the rhythmically moving left hand varied as a function of frequency as well as the point of discrete

---

[5] A fixed point is a point in phase space where velocity is zero. A limit cycle, as previously described, is a closed structure describing repetitive motion. For details, please see the chapter by Armin Fuchs in this volume.

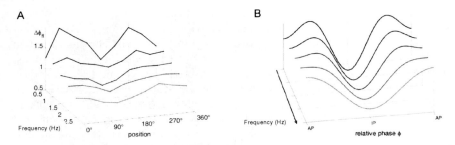

Fig. 6 A- **Normalized phase progression of the rhythmic hand movement during the discrete flexion movement**. AP = anti-phase (180°); IP = in-phase (0°). Values larger (smaller) than one indicate acceleration (deceleration) relative to uncoupled rhythmic movement. Note the change from a bimodal to unimodal structure with increasing frequency. B- **3D representation of the HKB model**. The potential describing the relative phase dynamics is plotted as a function of the frequency. At low frequencies, the system may, depending on the initial conditions, settle at the (global) minimum at IP or the (local) minimum at AP. As frequency increases, the local minimum at AP vanishes (the stable solution destabilized).

movement initiation. Figure 6 A summarizes our key result: it shows how at low frequencies the rhythmic hand accelerated when the discrete movement was initiated at around 90° and 270°, but not at 0° and 180°. When movement frequency increased, the acceleration vanished, and a deceleration was observed when the discrete movement was initiated at approximately 180° for high frequencies. This pattern shows a remarkable correspondence to the famous bimodal to unimodal probability distributions at increasing frequency in the rhythmic – rhythmic coordination scenario (compare figure 6 B). This finding is the first signature shadowing rhythmic–rhythmic coordination in the rhythmic—discrete case, and provides an exciting entry point to the claim that the HKB coupling can be invariant across components irrespective of their intrinsic dynamics.

In the framework of the dynamical patterns perspective, the interaction between discrete movements received even less attention. Kelso et al. (1979) explored this kind of coordination and evidenced as a main result that when the two hands performed simultaneously discrete actions with different amplitudes, the coupling induces a synchronization of the two segments due to the acceleration of the hand that performs the longer movement and a deceleration of the hand that performs the shorter movement.

In terms of modeling, the relevance of the HKB model to formalize the interaction between rhythmic movements is beyond dispute, but its application remains limited to this type of action and can by itself predict neither the dynamics of a discrete–discrete coordination nor that of a rhythmic–discrete coordination. It needs to understand and model the discrete movement, which has been done by Huys et al. (2008). Alternative models exist, for instance, Schöner (1986) proposed a model to capture the dynamics of coordination between two discrete movements using Gonzalez Piro oscillators coupled by the HKB coupling. However, this model was not validated experimentally, which limited its impact. Coordination between discrete and rhythmic movement was modeled by Sternad and collegues (2000) for the interaction between joints of the same limb and included a different ad hoc coupling function.

3 Extracting Principles Underlying the Formation of Coordinative Patterns: Convergence and Divergence

An apparent gap among the different paradigms on coordination exists. The rhythmic–rhythmic coordination is well, almost exhaustively, studied. More general forms of coordination involving non-rhythmical movements are far less studied and less well understood (see Sternad (2008) for an overview). It is evident that the mechanisms underlying human motor behavior in general and motor coordination in particular cannot be understood by a simple extrapolation of our knowledge (in term of concepts as well as tools) from the rhythmical paradigm. More precisely, concepts and tools suitable for investigation of rhythmic coordination are not simply convertible to other coordination. For example, the concept of stability developed for the rhythmic domain cannot be transferred to the rhythmic – discrete coordination, since the latter cannot be described by a fixed point dynamics or potential function; furthermore, discrete movements are by definition transient. Hence, what needs to be developed is a framework that is more general and includes the bimanual rhythmic description as a special case. Attempts thereof have been made (Schöner, 1990; Sternad, 2000), but have not provided any generalizable concepts so far.

To fill these gaps, a theoretical framework based on first principles was recently introduced in terms of phase flows Jirsa (2005). More precisely, these authors have recently proposed that mechanisms and processes responsible for movement generation, timing and motor coordination can be conceived as phase flows (see the Huys Chapter for detailed explanations and usage of phase flows). Phase flows describe the rate of change in a system's state space and govern the evolution of the system as a function of its current state. The fundamental aspect of phase flows is that they can be classified according to their flow topology. Any dynamic system with the same phase flow topology is member of the same class and essentially describes the same system. Then, the phase flow topology is an unbiased general representation of "processes" executed by a dynamical system within its phase space. Following the example of Schöner (1990), Jirsa & Kelso (2005) used the HKB coupling to couple two distinct phase flows capable of mimicking discrete movements. Later works established that these phase flows indeed are likely to be used by the human motor system (Fink et al., 2009; Huys et al., 2008).

In the current chapter, we wish to push this line of thought further and extend it to a generalizable conceptual framework. We summarize the nature of a biological coupling as follows: *If a given coupling between two or more effectors is weak, then the coupling will induce deformations in the phase flows of the effectors. In other words, the coupling will be considered weak, if the phase flow of the intrinsic dynamics of each subsystem is preserved as a whole (for instance, a limit cycle will remain a limit cycle rather than being turned into a bistable fixed point system by the coupling). The phase flow deformations by the coupling will be such that certain trajectories in the state space (equal to processes) will be favored and others will be suppressed.* In phase space, a selection of certain trajectories implies the exclusion of others; as a consequence the corresponding trajectories converge (motions attract each other) to each other toward the selected trajectory; or trajectories may equally diverge (motions repel each other) from each other in case of bistability in the system. The idea of convergence and divergence of phase space trajectories has already been put forward by Jirsa & Kelso (2005) when analyzing the dynamics of two coupled end effectors with discrete movement phase flows. More generally, Jirsa & Kelso (2004) suggested that convergence/divergence, integration/segregation or grouping/ungrouping of trajectories in phase space can be the key mechanisms responsible for trajectories formation, perceptual behavior and sensorimotor coordination. Observations from motor coordination domain revealed that interaction between limb movements (irrespective of the discrete or rhythmic nature of coordinated movement) implies that the movement of one accelerates or decelerates the other (and vice versa) (see Kelso et al., 1979 for discrete–discrete coordination, Calvin (2007) for rhythmic–discrete coordination, De Poel (2007) for rhythmic–rhythmic coordination). Our above formulation of the nature of coupling allows us to generalize this thought to arbitrary couplings and their function, as well as providing a precision of the vocabulary. This formulation of coupling includes the specific case of bimanual rhythmic coordination and can be generalized to phenomena including perception–action. In the following we wish to elaborate thereupon.

3.1 *Experimental Evidence for Convergence/Divergence in Perception–Action*

Studies on perception have showed that the formation of percepts change as a function of the spatiotemporal cohesion of the stimuli. Herzog & Koch (2001) showed that grouping can influences visual processes inducing the emergence of different visual percepts from the same stimuli. Furthermore, when listening to sequences of sounds, the sounds may be grouped together and so perceived as emanating from a single source, or perceived as separate auditory streams that originate from distinct sources. For understanding these auditory phenomena, Bregman (1990) introduced the concept of auditory streams as the critical phase in the perceptual process of interpreting the auditory environment. Auditory streaming entails two complementary mechanisms: 1/ how sounds cohere to form a sense of continuation is the subject of stream fusion or stream integration; 2/ since more than one source can sound concurrently, a second domain of study is how concurrent activities retain their independent identities — the subject of stream segregation. In normal hearing, the most obviously demonstrable cues for stream segregation are frequency separation and rate of presentation ((Bregman et al., 1971; Van Noorden, 1975). In an early work of Van Noorden (1975), listeners were presented with sequences of pure tones (A and B) differing in frequency and presented with different tempo. It was observed that when the difference in frequency between A and B was small and the tempo was low, sounds tended to be grouped together perceptually and listeners perceived a single melodic stream (one stream). In contrast, as the tempo and the frequency separation between alternating tones increase, so does the probability of hearing two distinct auditory streams (two streams). In this case, listeners perceived the sequence as consisting of two streams, one formed by the A tones and the other by the B tones. This experiment revealed that different perceptual patterns emerge when manipulating two parameters: the frequency difference between tones and the tempo (tone repetition) that act as control parameters inducing changes in percept dynamics. In the context of our present chapter, two patterns or modes of perception emerge, that is segregation (two streams) and integration (one stream). Both perceptual modes are in competition with each other (see Almonte et al., 2005 for a dynamic modeling study using competition networks), displaying either monostability (only one mode exists) or bistability (both modes are possible) as a function of the control parameters, not unlike the foreground–background separation in vision or the situation in bimanual coordination as discussed previously. In a conceptually related study on multisensory integration, Dhamala et al. (2007) used functional magnetic resonance imaging (fMRI) to shed light on the neural mechanisms underlying perceptual integration and segregation. They presented two stimuli (auditory stimuli precede the visual one and vice versa) together with a time delay separating them (Δt from 0 to 200 ms). The control parameters of stimulation frequency and time delay were varied and subjects were required to classify percepts. As a function of the timing parameters (control parameters), three percepts were identified: one "synchronous" percept (the tone and the flash were perceived simultaneously), two "segregated" percepts, one corresponding to sound preceding the light and the other light preceding the sound; and one "drift" percept corresponding to a perception, in which the subjects were unable to report the order. These findings revealed the existence of two distinct perceptual states, again a bistable regime composed of both integrated and segregated patterns and an intermediate regime (drift) where none of the segregated or integrated patterns were clearly adopted and maintained. The analysis of the fMRI data identified neural networks involving prefrontal, auditory, visual, and parietal cortices and midbrain regions for perceptions of physically integrated, segregated (auditory–visual or visual–auditory) events and drift. Remarkably, the perception of integration and segregation revealed differential network activations, whereas the drift network was mostly left with rudimentary network activations. These findings provide further, and in addition neural, support for the ontological interpretation of integration and segregation as behavioral (perceptual) patterns of operation.

We wish to emphasize one more time that perceptual patterns, integration and segregation, correspond to an emerging mode of operation and result in a stable (perceptual) state displaying all properties (multistability, hysteresis, transitions, etc.) known from dynamic systems and as discussed previously. Hence, we have to conclude that the corresponding state space, in which the perceptual dynamics evolves (and which is so far inaccessible to us by measurement), has converging and diverging trajectories; hereby the converging trajectories evolve toward the stable patterns of integration and segregation. Then the diverging trajectories correspond to the areas of phase space in which the flows evolve into opposite directions. This reasoning allows us to better formalize the concepts of convergence and divergence, as well as segregation and integration. Convergence and divergence describe the directions of flow in the phase space, whereas integration and segregation refer to modes of operation of the perception action system. This is in subtle contrast to the formulation by (Kelso, 1995; Lagarde & Kelso, 2006), in which "integration" refers to grouping of individual coordinating elements that lead to the emergence of phase- and frequency-locked (stable) coordinative behaviors, whereas "segregation" is clearly associated with the notion of loss of frequency and phase locking — as in phase drift for example — induced by the fact that components become independent. The difference is mostly of semantic nature, though it needs to be clarified here. Lagarde & Kelso (2006) interpret segregation as the loss of integration, which is closer to our here formalized notion of phase flow divergence.

Obviously, evidence for binding, grouping and integration is not restricted to the study of perception and has been studied in detail in various works of sensorimotor coordination as already discussed in this chapter. Other relevant works include the one by Kelso et al. (1990) who reported that the coordination pattern of synchronizing the index finger flexion with an auditory metronome signal was significantly more stable than the syncopation pattern where the flexion of the index finger falls between two successive auditory events. These sensorimotor patterns correspond directly to the notions of integration (synchronization) and segregation (syncopation), as we presented them here. These results have been confirmed in experiments dealing with multisensory coordination dynamics (Kelso et al., 2001; Lagarde & Kelso, 2006). Indeed, the authors showed that the organization of movement in harmony with different stimuli from different sensory modalities (audition and tact) was correlated to their temporal and spatial congruency. For example, in Kelso et al's experiments (2001) subjects were asked to coordinate finger flexion or extension with an auditory metronome, the frequency of which was systematically increased. At the same time, they were required to touch a haptic contact realized by a physical stop, located either in coincidence with or in counterphase with the auditory stimulus. Results showed that at a critical frequency, irrespective of whether subjects used flexion or extension for synchronizing with the (auditory) metronome, the subjects (re)organized their coordination dynamics of sound, touch and movement. At low frequencies regimes, the two separate modalities were segregated; however when frequency increased, stable coordination was maintained only if auditory and haptic signals were integrated to compose a *"coherent perception–action unit"* Kelso et al. (2001).

3.2 Formalization of Convergence/Divergence in the State Space

As shown above, the phase flow in state space (or phase space) contains all information about the dynamics of a given system. Since the phase flow topology is the only determining criterion about the nature of a system, the invariant flow elements in phase space giving rise to topologically relevant structures will be of particular interest. The latter includes fixed points, limit cycles and separatrices, where a separatrix is defined as a set of phase space points at which the direction of the phase flow vector changes its orientation (in a given dimension) pointing away from the separatrix. Stated simply, a separatrix is an element that "separates" two regions of phase space with opposite flows. In a one-dimensional system, every stable fixed point is an area of convergence; every unstable fixed point is an area of divergence and identical to a separatrix. In two-dimensional systems stable fixed points remain to be the areas of convergence; however, the saddle points and separatrices act as areas of divergence, but not the unstable fixed points (though the latter are topologically relevant). In the following, we illustrate these notions with various examples.

Our *one-dimensional system* of choice is the HKB system, but not in its potential representation, but rather formalized by its phase flow in state space. It reads as follows:

$$\dot{\phi} = f(\phi) = -a\sin\phi - 2b\sin 2\phi$$

with ϕ as the relative phase between two nonlinear oscillators. The phase flow $f(\phi)$ is a vector field on the line and prescribes the rate of change $\dot{\phi}$ at each value of ϕ. When $\dot{\phi} > 0$ the vector field points to the right and when $\dot{\phi} < 0$ the vector field points to the left. When $\dot{\phi} = 0$, there is a fixed point. The fixed point is stable when the phase flow points toward it, else it is unstable. The phase flow for various values of a and b is plotted in figure 10. If the two oscillators are uncoupled, $a=b=0$, then all phase relationships are possible, i.e., ϕ can take any value. Thus $\dot{\phi} = 0$ for all values of ϕ. These fixed points are neither stable nor unstable. If a perturbation is imposed on one of the two oscillators or to the both, it induces deviations of the trajectories of the oscillators and consequently the establishment of new phase relationships. If the two oscillators are coupled, then their dynamics is described in figure 10. When $\frac{|b|}{|a|} = 1$ and $a,b>0$ (figure 7 A), then we find the fixed points ϕ^* by setting $\dot{\phi}=0$ and obtain $\phi^* = \pm\frac{\pi}{2}; \pm\pi; \pm\frac{3\pi}{2}$. To determine the stability of these fixed points, let us examine the vector field: when $-a\sin\phi - 2b\sin 2\phi > 0$, the flow is to the right and it is to the left when $-a\sin\phi - 2b\sin 2\phi < 0$. Thus, $\phi^* = \pm\pi; 0$ are stable fixed points, since the flow converges toward them. These points act as areas of convergence, where $\phi^* = 0$ indicates integration (the oscillators move synchronously) and $\phi^* = \pm\pi$ indicates segregation (the oscillators move in anti-phase). On the contrary, $\phi^* = \pm\frac{\pi}{2}; \pm\frac{3\pi}{2}$ are unstable fixed points and their neighboring flow points away from them. These points are areas of divergence. The arrows represent the vector field: when starting at $\pi/4 < \phi_0 \leq \pi/2$, the flow moves to the left increasingly slowly until it crosses $\phi = \pi/4$ where $f(\phi)$ reaches its minimum. Then the flow starts moving increasingly faster when approaching the stable fixed point $\phi=0$. Starting at $\pi/2 < \phi_0 \leq 2\pi/3$, the flow moves to the right faster and faster until it crosses $\phi = 2\pi/3$. Then the flow starts slowing down and approaches the stable fixed point $\phi=\pi$. The velocity with which the phase is attracted by the fixed point located at $\phi=0$ is greater than that at $\phi=\pi$. In terms of behavior, it reflects the bistability of the coordination dynamics at low frequency regime: indeed in-phase and anti-phase patterns are the two attractors of the coordinative behavior and the in-phase pattern is more stable than the anti-phase pattern. As the

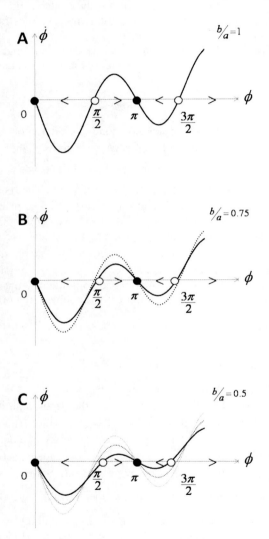

Fig. 7 Phase flow representation of the dynamics of the relative phase (φ) obtained for different values of b/a (see text for explanation).

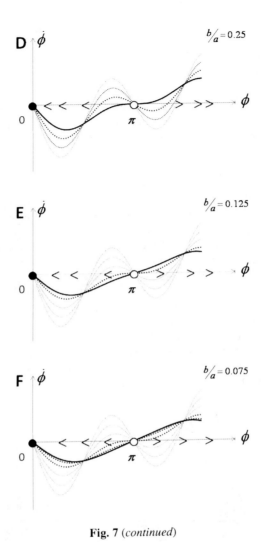

Fig. 7 (*continued*)

movement frequency increases, the ratio decreases, $\frac{|b|}{|a|} < 1$. Note that $\frac{|b|}{|a|}$ expresses the relative importance of the phase attraction at 0 and π; when $\frac{|b|}{|a|} = 0.75$, the shape of the phase flow changes but the phase flow topology stays the same (figure 7 B, C). Indeed, all stable fixed points are still located at the same locations, but the unstable fixed points shift progressively from $\phi^* = \pm\frac{\pi}{2}; \pm\frac{3\pi}{2}$ to $\phi = \pm\pi$. When $\frac{|b|}{|a|} \geq 0.25$, the topology of the phase flow changes qualitatively (Figure 7 D, E, F). The unstable fixed points collapse with the stable fixed points at $\phi = \pm\pi$ and leave the system with one unstable fixed point only. The slope of $f(\phi)$ is negative at $\phi=0$, signaling the presence of a stable fixed point and is positive at $\phi=\pi$ indicating the presence of an unstable fixed point. In terms of behavior, it corresponds to the fact that when the frequency increases, the anti-phase pattern no longer exists and a phase transition occurs to the more stable in-phase pattern.

For the *higher-dimensional system*, we represent indeed the same HKB dynamics, but in its original four-dimensional configuration (two coupled oscillators). We start off with figure 2, representing the two oscillators in their respective two-dimensional phase planes. In conjunction, these two phase planes span a four-dimensional system, in which the system as a whole is characterized by one limit cycle. The two phase planes then represent projections into the individual subspaces. If they are uncoupled, then all phase relationships are possible corresponding to different angles of the trajectory in the four-dimensional space. A representation thereof is found in figure 8, in which three axes (position x_1 and x_2 of the two oscillators and velocity y_1) are shown. The different trajectories with different relative phases cover continuously a cylinder and no particular phase is preferred.

When the oscillators are coupled, then the trajectories corresponding to the relative phase 0 and 180° are selected and stabilized. The coupling deforms the uniform phase flow seen in figure 8 and causes the trajectories to converge to the two stabilized patterns of in-phase and antiphase (see figure 9).

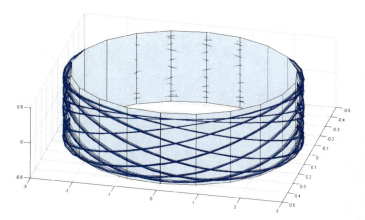

Fig. 8 Trajectories of the same oscillators are represented in a 3D phase space. A cylinder is plotted along with the trajectories to help their visualization. When the two oscillators are uncoupled, they are able to establish all phase relationships, and the trajectories corresponding to these relationships are represented along the cylindrical manifold.

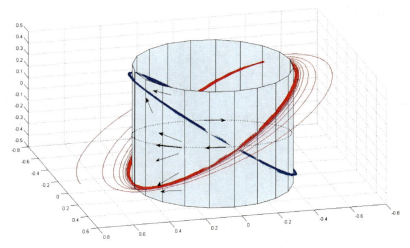

Fig. 9 When the two oscillators are coupled, the coupling restricts the phase relationships that the two oscillators can establish, only two persist: one trajectory corresponding to the in-phase (0°) in blue and another corresponding to anti-phase (180°) motion in red are shown. These trajectories indicate regions of convergence. The arrows indicate the newly established phase flow. In the middle on the cylinder (black line) is the separatrix, where the phase flow diverges toward either of the convergence zones.

As the control parameter passes through the critical value, the phase flow reorganizes and the convergence zone around the anti-phase trajectory disappears, rendering itself to a zone of divergence (see figure 10).

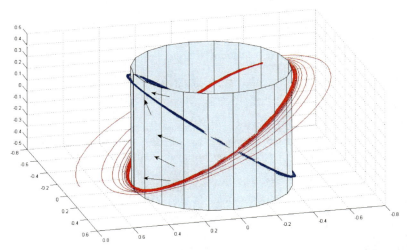

Fig. 10 The same situation is shown as in figure 9, but for a value of the control parameter greater than the critical value. Only the in-phase trajectory (blue) is stable, the anti-phase trajectory (red) is unstable.

4 Conclusions

In the current chapter, we have discussed in great detail the effects of coupling between biological oscillators with a focus on movement coordination. We attempted to provide an overview of the existing knowledge on dynamic descriptions of movement phenomena. We particularly emphasized the perspective of phase flows, in which biological coupling can be understood as weak deformations of the phase flow of the intrinsic dynamics of biological oscillators. The line of thinking provides us with a generalized description of coordination dynamics with the potential of applications to non-rhythmic behaviors. Future research will show its fruitfulness in applications to behavioral neuroscience.

References

1. Almonte, F., Jirsa, V.K., Large, E.W., Tuller, B.: Integration and segregation in auditory streaming. Physica D 212, 137–159 (2005)
2. Anderson, P.W.: More is different. Science 177, 393–396 (1972)
3. Bernstein, N.A.: The Co-ordination and Regulation of Movements. Pergamon Press, Oxford (1967)
4. Bregman, A.S., Campbell, J.: Primary auditory stream segregation and perception of order in rapid sequences of tones. Journal of Experimental Psychology: Human Perception & Performance 89, 244–249 (1971)
5. Bregman, A.S.: Auditory scene analysis. MIT Press, Cambridge (1990)
6. Calvin, S., Huys, R., Jirsa, V.: Coordination dynamics of discrete and rhythmic bimanual movements. In: European Workshop On Movement Science, Mechanics, Physiology, Psychology, pp. 114–115. Verlag Sport und Buch Strauss, Köln (2007)
7. Calvin, S., Huys, R., Jirsa, V.: Interference effects in bimanual coordination are independent of movement type. Journal of Experimental Psychology: Human Perception & Performance (in press) (2010)
8. De Poel, H.J., Peper, C.E., Beek, P.J.: Handedeness-related asymmetry in coupling strength in bimanual coordination: Furthering theory and evidence. Acta Psychologica 124, 209–237 (2007)
9. Dhamala, M., Assisi, C.G., Jirsa, V.K., Steinberg, F.L., Kelso, J.A.S.: Multisensory integration for timing engages different brain networks. Neuroimage 34, 764–773 (2007)
10. Fink, P.W., Kelso, J.A., Jirsa, V.K.: Perturbation-induced false starts as a test of the Jirsa-Kelso excitatory model. Journal of Motor Behavior 41, 147–157 (2009)
11. Haken, H.: Synergetics: an introduction. Springer, Heidelberg (1977)
12. Haken, H.: Advanced synergetics. Springer, Heidelberg (1983)
13. Haken, H., Kelso, J.A.S., Bunz, H.: A theoretical model of phase transitions in human hand movements. Biological Cybernetics 51, 347–356 (1985)
14. Herzog, M.H., Koch, C.: Seeing properties of an invisible object: Feature inheritance and shine-through. Proceedings of the National Academy of Sciences 98, 4271–4275 (2001)
15. Huys, R., Studenka, B.E., Rheame, N., Zelaznik, H.N., Jirsa, V.K.: Distinct timing mechanisms produce discrete and continuous movements. PLoS Computational Biology 4, (e1000061) (2008)

16. Jirsa, V.K., Kelso, J.A.S.: Coordination dynamics: Issues and trends. Springer, Heidelberg (2004)
17. Jirsa, V.K., Kelso, J.A.S.: The excitator as a minimal model for the coordination dynamics of discrete and rhythmic movement generation. Journal of Motor Behavior 37, 35–51 (2005)
18. Kelso, J.A.S.: On the oscillatory basis of movement. Bulletin of the psychonomic society 18, 63 (1981)
19. Kelso, J.A.S., Jeka, J.: Symmetry breaking dynamics of human multilimb coordination. Journal of Experimental Psychology: Human Perception & Performance 18, 645–668 (1992)
20. Kelso, J.A.S., Buchanan, J.J., Wallace, S.A.: Order parameters for the neural organization of single, multijoint limb movement patterns. Experimental Brain Research 85, 432–444 (1991)
21. Kelso, J.A.S.: Phase transitions and critical behavior in human bimanual coordination. American Journal of Physiology 246, 1000–1004 (1984)
22. Kelso, J.A.S.: Dynamic Patterns. The Self-Organization of Brain and Behavior. MIT Press, Cambridge (1995)
23. Kelso, J.A.S., Delcolle, J.D., Schöner, G.: Action-perception as a pattern formation process. In: Jeannerod, M. (ed.) Attention and Performance XIII, pp. 139–169. Erlbaum, Hillsdale (1990)
24. Kelso, J.A.S., et al.: Haptic information stabilizes and destabilizes coordination dynamics. Proceedinsg of the Royal Society London B 268, 1207–1213 (2001)
25. Kelso, J.A.S., Southard, D.L., Goodman, D.: On the coordination of two-handed movements. Journal of Experimental Psychology: Human Perception & Performance 5, 229–238 (1979)
26. Lagarde, J., Kelso, J.A.S.: Binding of movement, sound and touch: multimodal coordination dynamics. Experimental Brain Research 173, 673–688 (2006)
27. Latash, M.: Modulation of simple reaction time on the background of an oscillatory action: implications for synergy organization. Experimental Brain Research 131, 85–100 (2000)
28. Michaels, C.F., Bongers, R.M.: The dependence of discrete movements on rhythmic movements: simple RT during oscillatory tracking. Human Movement Science 13, 473–493 (1994)
29. Perko, L.: Differential equations and dynamical systems. Springer, New York (1991)
30. Salesse, R., Temprado, J.J., Swinnen, S.P.: Interaction of neuromuscular, spatial and visual constraints on hand-foot coordination dynamics. Human Movement Science 24, 66–80 (2005)
31. Schöner, G., Haken, H., Kelso, J.A.S.: A stochastic theory of phase transitions in human hand movement. Biological Cybernetics 53, 442–452 (1986)
32. Schöner, G.: A dynamic theory of coordination of discrete movement. Biological Cybernetics 63, 257–270 (1990)
33. Sternad, D.: Rhythmic and discrete movements-behavioral, modeling and imaging results. In: Fuchs, A., Jirsa, V.K. (eds.) Coordination: Neural, Behavioral and Social Coordination, pp. 105–136. Springer, Heidelberg (2008)
34. Sternad, D., Wei, K., Diedrichsen, J., Ivry, R.B.: Intermanual interactions during initiation and production of rhythmic and discrete movements in individuals lacking a corpus callosum. Experimental Brain Research 176, 559–574 (2007)

35. Sternad, D., Dean, W.J., Schaal, S.: Interaction of rhythmic and discrete pattern generators in single-joint movements. Human Movement Science 19, 627–664 (2000)
36. Sternad, D., Wei, K., Diedrichsen, J., Ivry, R.B.: Intermanual interactions during initiation and production of rhythmic and discrete movements in individuals lacking a corpus callosum. Experimental Brain Research 176, 559–574 (2007)
37. Strogatz, S.H.: Non linear dynamics and chaos with applications to physics, biology, chemistry, and engineering. Perseus Books Publishing LLC, Cambridge (1994)
38. Tuller, B., Kelso, J.A.S.: Environmentally-specified patterns of movement coordination in normal and split-brain subjects. Experimental Brain Research 75. 306–316 (1989)
39. Turvey, M.: Coordination. American Psychologist 45, 938–953 (1990)
40. Van Noorden, L.P.A.S.: Temporal coherence in the perception of tone sequences. Unpublished doctoral dissertation, Eindhoven University of Technology (1975)
41. Von Holst, E.: The behavioral physiology of animals and man. Methuen, London (1973)
42. Wei, K., Wertman, G., Sternad, D.: Interactions between rhythmic and discrete components in a bimanual task. Motor Control 7, 134–154 (2003)
43. Wimmers, R.H., Beek, P.J., Van Wieringen, P.C.W.: Phase transitions in rhythmic tracking movements: A case of unilateral coupling. Human Movement Science 11, 217–226 (1992)
44. Yamanishi, J.I., Kawato, M., Suzuki, R.: Studies on human finger tapping neural networks by phase-transition curves. Biological Cybernetics 33, 199–208 (1979)

Do We Need Internal Models for Movement Control?

Frédéric Danion

Abstract. The issue of how humans and animals perform accurate movements has been addressed in various ways. Although this book is promoting concepts stemming from dynamical systems theory, other approaches have contributed to the understanding of movement as well. Among others, the equilibrium point theory and the computational theory deserve to be listed for their contribution to this field of research called motor control. In this chapter, using single-joint rhythmic movement as an example, I will start first emphasizing the respective contributions and drawbacks of each approach. Then I will address the issue of parameter selection. Indeed, despite diverging opinions about the possible nature of control parameter(s), all three approaches must deal with the problem of how adequate parameter(s) to achieve a desired movement are selected. At the end of this chapter, I will expose how the concept of internal model may offer a solution to this problem.

1 Dynamical Systems Theory

Dynamical systems theory and its applications have been addressed extensively in the other chapters of this book. In this section, I will concentrate on a number of studies issued from this approach that have addressed the control of single-limb rhythmic movements (see Huys, this volume). As reported by Beek and colleagues (1996), in those studies, the dynamics of rhythmic movement are typically captured by a second order differential equation of the form:

$$m\ddot{x} + f(x,\dot{x})\dot{x} + g(x) = 0 \qquad (1)$$

where x is the angular displacement, $\dot{x}$ and $\ddot{x}$ the first and second derivative of x with respect to time. The first term expresses the inertia of the system, the second

---

Frédéric Danion
Université de la Méditerranée, UMR 6233 "Movement Science Institute", CNRS,
Faculté des Sciences du Sport, 13288, Marseille cedex 09, France

expresses the system's damping (viscosity), and the third one accounts for the system's stiffness (elasticity).

Several ways to implement damping and stiffness have been proposed to account for experimental data. In a seminal study, Kay and colleagues (1987) investigated oscillatory wrist movement at various frequencies (from 1 to 6 Hz). The authors emphasized that wrist movement kinematics can be well accounted by the following differential equation:

$$\ddot{x} + (\alpha + \beta \dot{x}^2 + \gamma x^2)\dot{x} + \omega x^2 = 0 \qquad (2)$$

In this equation, stiffness is a linear function while damping is a non-linear function. For those familiar with oscillators, equation 2 is in fact a combination of Rayleigh (when $\gamma=0$) and Van Der Pol oscillators (when $\beta=0$). These two types of oscillators exhibit limit cycles and self-sustained oscillations. The interesting feature of this so-called hybrid oscillator is that by adjusting the value of parameters α, β, γ, and ω, it becomes possible to account for the kinematics of wrist movements over a wide range of frequencies. Rhythmic movements are not perfectly sinusoidal. In fact, there are subtle changes in their kinematics as a function of frequency (Kay et al., 1987; Beek et al., 1996). By adjusting the value of each parameter, it is possible to account for a wide variety of movement kinematics. For instance, to vary the frequency of oscillation, one can change the value of ω. Then, by modulating the damping terms (α, β, γ), it becomes possible to make the oscillations more or less harmonic. In addition, the amplitude of movement is (directly or indirectly) under the influence of all parameters.

In another study, performed by Mottet and Bootsma (1999), a differential equation was also proposed to account for the changes in movement kinematics resulting from changes in accuracy constraints during a reciprocal aiming task. The equation was the following one:

$$\ddot{x} + Ax - Bx^3 - Cx + D\dot{x}^3 = 0 \qquad (3)$$

While keeping the same equation, and adjusting the value of the parameters (A, B, C, and D), the authors obtained a good fit of experimental trajectories performed under various indices of difficulty (by manipulating target width and movement amplitude). The corresponding set of parameters for each index of difficulty is provided in Figure 1.

The earlier studies show the power of differential equations to account for rhythmic movements. However, one may eventually question the physical or biological relevance of the parameters present in those equations (Rosenbaum, 1998). Although some of the terms are often proposed as the expression of damping and stiffness present into the oscillating system, should we understand that ω in *equation 2* really accounts for the stiffness of the wrist? The answer is no because, even if some salient properties of an oscillating system can be derived from its kinematics during steady-state regimes, the only valid way to assess its stiffness is to investigate its response to a mechanical perturbation (the smaller the effect, the greater the stiffness). This problem is illustrated by the

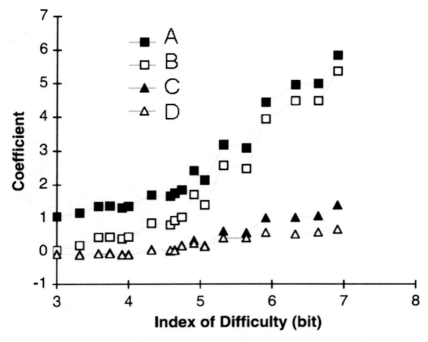

Fig. 1 Value of coefficients A, B, C, and D, as a function of the index of difficulty. Increasing difficulty shows a parallel increase of the damping (triangles) and stiffness (squares) coefficients, denoting an increasing contribution of the nonlinear terms. (Reprinted with permission from Mottet & Bootsma, 1999)

following example. Each of us can maintain a slow rhythmic movement of the wrist while modulating concurrently the degree of cocontraction between agonist and antagonist muscles. Although the kinematics remain unaffected by this procedure, the response to a given mechanical perturbation will change depending on the level of muscle cocontraction. With a higher level of cocontraction, the deviation from the initial trajectory will decrease, thereby supporting a higher stiffness. This example demonstrates that observations of phase portraits are not sufficient to infer the 'true' stiffness of an oscillating system, i.e. its resistance to perturbation.

Another difficulty arises from the fact that it is hard to compare values of stiffness measured through perturbation techniques, with values of stiffness estimated in the earlier studies, simply because they are provided in completely different units. For instance, stiffness is expressed in Nm/rad in Bennet et al. (1992), while it is expressed in Hz in Kay et al. (1987). This partly explains why discrepancies can be found between the two approaches. For instance, although techniques based on mechanical perturbation indicate that elbow stiffness is constantly modulated during each cycle of movement (Bennet et al., 1992), some studies stemming from the dynamical approach assumed that stiffness is not time-varying (e.g. see equation 2).

Altogether, should those conflicting interpretations question the biological relevance of the parameters used in differential equations? Or did we simply misunderstand some of the concepts behind those differential equations? A response can partly be found in the paper of Beek et al (1996), who state that in the dynamical systems approach, *"stiffness and damping are abstract control parameters that refer to the space-time behavior of the system as a whole, whereas in (bio)mechanics, they refer to locally identifiable concrete force and structures"* (p.1077). This means that stiffness and damping do not account only for the physical properties of the limb, but also include the dynamical properties stemming from the nervous system (at both cortical and spinal levels), as well as the physical properties of the environment in which movement is taking place (external forces). This means that if the subject is oscillating vertically his or her wrist within a stiff and/or viscous manipulandum, the resulting set of parameters needed to account for the wrist kinematics refers to both the physical properties of the limb, the nervous system, the apparatus, and the external forces (here gravity). The dynamical approach, in general, does not evaluate the relative contribution of each of those components. The strategy is rather to propose a single equation that summarizes all those contributions.

Obviously, when we perform an oscillatory movement, we have the possibility to modulate several of its features. The rate at which we oscillate the limb (frequency) and the distance covered between two extrema (amplitude) can be adjusted within a certain range. Let's assume that you are oscillating your limb at 1 Hz, and that now you want to oscillate at 2 Hz. What are you supposed to do? Assuming that our earlier differential equation is still adequate for this situation, something in your brain must happen to change the value of one or several parameters, otherwise your movement will not change. Returning to the study of Mottet & Bootsma (1999), there are several possibilities to increase frequency. A first possibility would be to decrease C. A second option would be to increase A. A third possibility is to do both. Just by using this simple example, we already see that there are multiple options to obtain the desired effect. This well-identified problem in motor control is known as the redundancy problem and has been first pointed out by Bernstein (1967) to emphasize the difficulty of controlling the end-effector in multi-joint movement. Despite this ambiguity in selecting a strategy to increase movement frequency, none of us would end up decreasing his or her frequency, even if suddenly asked. This observation simply means that the brain has a priori some knowledge about the influence of those parameters on movement kinematics. How this knowledge is used and stored in the brain will be addressed in section 5.

2 Positional Control and Equilibrium Point Theory

In contrast to the dynamical systems approach, the equilibrium point theory allows to disentangle the relative contribution of external forces, limb properties, and motor commands. A key difference is that this theory suggests that movement is initiated through changes in parameters that have a biological meaning. In its most

Do We Need Internal Models for Movement Control?

famous version, this parameter corresponds to the threshold of the tonic stretch reflex (for reviews see Feldman 1986; Latash, 1993; Feldman & Levin, 1995). More specifically, it is assumed that for each muscle, the brain can adjust the threshold length (λ) at which the tonic stretch reflex is engaged. If the actual length L of a muscle is greater than λ, then the tonic stretch reflex is elicited, and a resisting force proportional to L minus λ is generated by the muscle, meaning that electromyographic (EMG) activity should be observed. In contrast if L is equal or inferior to λ there is no EMG activity and the muscle resisting force is null. Altogether, here muscles are viewed as a special kind of springs. First, they can generate force only in one direction (so as to shorten). Second, their resting length is adjustable. A crucial point is that, even if the λ of the biceps is held constant, EMG activity in my biceps can change depending on the conditions. For instance, if there is no external load applied on my biceps, then $L = \lambda$, and my biceps is at rest (i.e. no EMG activity). In contrast, if an external load (e.g. gravity) stretches my biceps beyond its resting length, EMG activity is instantiated to resist this external load. It is fundamental to realize that in this framework, EMG activity and movement are only indirectly connected with motor commands, since EMG emerges from the interaction between motor commands and external forces. Therefore muscle activity (and muscle force) must be interpreted cautiously, as well as differences across EMG patterns obtained under different types of external load.

Obviously, we have more than one muscle, and this ultimately requires the existence of multiple λs. Instead of dealing with one λ per muscle, many defenders of the equilibrium point theory assumed that muscles being functionally identical are driven by a single λ (Latash, 1993; Feldman & Levin, 1995). In the case of muscles acting at the same joint, this lead to one λ for the agonist muscle group, and one λ for the antagonist muscle group. The difference and the sum between the two λ define new variables that still have physical meaning. If we stick to the spring analogy, by changing the resting lengths of the springs in opposite directions, we modify the equilibrium position of the joint. Now if we modify the resting length of the springs in similar direction, we do not affect the equilibrium position of the joint, but we change its stiffness. If λ for agonist and antagonist are shortened by a similar amount, this leads to increased joint stiffness. Searching for simplification, a new set of variables has been proposed. The reciprocal variable (R) accounts for changes of λ in opposite direction, while the coactivation variable (C) accounts for changes of λ in similar direction. This change in coordinate system is ensured by the following equations:

$$R = \left(\lambda_{ag} - \lambda_{ant}\right)/2 \text{ and } C = \left(\lambda_{ag} + \lambda_{ant}\right)/2 \qquad (4)$$

with λ_{ag} and λ_{ant} accounting for the length of the agonist and antagonist muscle above which a tonic stretch reflex is engaged. If λ_{ag} and λ_{ant} refer to threshold joint angles, the sign + and − should be exchanged. Note that the existence of separate neuronal systems for reciprocal activation and coactivation is supported by cortical recordings in monkeys (Humphrey and Reed, 1983).

To account for single-joint oscillatory movement, Feldman (1980) suggested that beyond 1 Hz, R shifts back and forth by while C is held relatively constant between certain limits (see Figure 2A). Although more complex changes in R and C commands (see Figure 2B,C) have been proposed later by Latash (1992), in both versions, the C command is envisaged to increase as a function of movement frequency (this idea being reminiscent from what we proposed under the dynamical system approach). In addition, both versions suggest that the amplitude of shifts in R command is tailored so as to match the desired movement amplitude. Similarly, the frequency of shifts in R command is also tailored to match the desired frequency of movement.

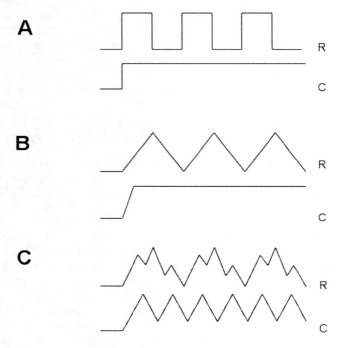

Fig. 2 Patterns of R and C commands hypothesized for the maintenance of a rhythmic movement.

Let us now raise a few issues showing why controlling movement through shifts in equilibrium point does not necessarily simplify movement control. When I exposed the influence of the R and C commands, I did so in the absence of external forces. However, the net effect of changes in R and C commands becomes less trivial in the presence of an external load (Gottlieb, 1995; Flanagan et al., 1995). For instance, if we hold our forearm stable horizontally (i.e. against gravity), a change in R command will affect the equilibrium position of the limb, meaning that the forearm will move up or down (depending on the change in R). On the other hand, one should realize that a change in C command can also move the forearm upward or downward; for instance, if C drops, the forearm will move down. As a result, the same shift in elbow position can a priori be obtained through many changes in control parameters, making the interplay between R and C commands more complex than initially envisaged. Again, we are facing the redundancy problem in selecting the appropriate set of motor commands. At last, because movement emerges from the interaction between R, C, and the external forces, the issue of maintaining the same kinematics against different external loads is not trivial. As pointed out by Gottlieb (1998a) in a commentary, the rules for changes in control parameters are neither obvious nor simple. It follows that the issue of how the brain sends the adequate motor signals to the muscles to generate the desired movement has been rarely addressed by the promoters of the equilibrium point theory (although see Gribble & Ostry, 2000), but I will come back to this point later.

3 Force Control and Computational Theory

In contrast to the dynamical system approach and equilibrium point theory, the computational theory proposes that the "strings" pulled by the brain to generate movements are muscle forces/torques. Because EMG reflects neural activation, and because EMG relates to muscle force (Gottlieb & Agarwal, 1971; Olney & Winter, 1985; Thelen et al, 1994), some researchers find attractive that the nervous system could deal directly with muscle force when producing voluntary movements. This view is supported by numerous studies in which invariant strategies were found when investigating changes in EMG activity as a function of task parameters such as movement amplitude, or movement speed (Gottlieb et al, 1989; Almeida et al, 1995; Gottlieb, 1998b). For instance, in Figure 3, one can quickly notice that changes in EMG are relatively simple when a fast movement is performed over different amplitudes. Altogether, modulating the amplitude of the agonist burst (i.e. biceps) and the timing of the antagonist burst (i.e. triceps) provides a strategy that can account for a large variety of movements. This interest in EMG signals is also encouraged by many neurophysiological studies showing close relationships between the activity of cortical neurons and muscle force in the monkey (Georgopoulos et al, 1992; for a review see Ashe, 1997). In humans, using brain imaging techniques, Dai and colleagues (2001) have found that hand muscle activation (based on EMG and grip force) was directly proportional to the amplitude of the brain signal.

Controlling directly muscle force may seem like an attractive way to produce movements, however, this may rapidly become a difficult task. Indeed, although the relationship between EMG and muscle force is relatively linear in static (isometric) conditions (Lippold, 1952), it is no longer the case during movement, for at least three reasons. First, muscles act on joints by means of torques, and moment arm of muscles can vary in a rather complex fashion as function of joint angle. Additionally, those functions can be relatively different across muscles (Pigeon et al, 1996). Second, even when considering muscle force only, it has been demonstrated that the ability of a muscle to produce force changes as a

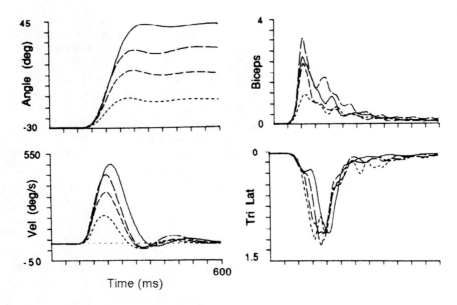

Fig. 3 Kinematics and filtered EMG patterns associated with fast elbow joint flexion movements performed over different amplitudes (adapted with permission from Gottlieb et al, 1989)

function of its length, as well as the rate of change of its length. Again those functions, known as force-length and force-speed relations, are both non linear (Hill, 1938). Third, even in the simplest case (i.e. when muscle length is constant, isometric conditions), progressive increase in non-linearity is evident in EMG-force relationship above 50% of maximal force (Solomonow et al, 1986). Altogether, there are many evidences that the relationship between neural activation and muscle force/torque is not trivial.

On top of non-linearities between neural activation and muscle force, recall that movement results from the interaction of many forces, muscle force being just one of them. Indeed, in many circumstances other forces, such as gravitational force, coriolis force, interaction torques, reactive force, contribute to movement as well. As a result, if muscle forces are explicitly controlled by the nervous system, they must be precisely tailored to achieve the desired trajectory. Lets examine the different forces/torques that contribute to the motion of the arm involving the shoulder and the elbow joint. The set of equations shown in Figure 4 allows for computing muscle torques based on individual joint kinematics. The method used here is called inverse dynamics. The 'philosophy' behind this method is to calculate the net torque acting at each joint (this is done based on the acceleration of the limb and its inertia) and to subtract for each value all the non-muscular torques also known to contribute to movement. For instance in our case, we know that the movement of the shoulder joint influences the movement of the

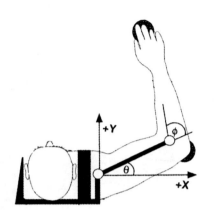

Elbow joint torques

Interaction = $-\ddot{\theta}[A \cos(\phi) + B] - \dot{\theta}^2 A \sin(\phi)$

Net = $\ddot{\phi} B$

Muscle$_{elbow}$ = Net − Interaction

Shoulder Joint Torques

Interaction = $-\ddot{\phi} A \cos(\phi) + (\dot{\phi} + \dot{\theta})^2 A \sin(\phi)$

Net = $\ddot{\theta}[C + A \cos(\phi)]$

Muscle$_{shoulder}$ = Net − Interaction − Muscle$_{elbow}$

Symbols

$A = m_2 L_1 r_2 + m_d L_d r_d$

$B = I_2 + m_2 r_2^2 + I_d + m_d r_d^2$

$C = I_1 + m_1 r_1^2 + (m_2 + m_d) L_1^2$

where m is mass, r is distance to center of mass from proximal joint, L is length, I is inertia, θ is shoulder angle, and ϕ is elbow angle. The subscripts are defined as follows: 1 is upper arm segment, 2 is forearm/hand segment, and d is air sled device.

Fig. 4 Set of equations accounting for the movement of a bi-articular movement performed in the horizontal plane (adapted with permission from Sainburg & Kalakanis, 2000).

elbow joint by means of interaction torques. As a result, these interaction torques are subtracted from the net torque of the elbow. When all non-muscular contributions are removed, the leftover is considered as the joint muscle torque. Although this set of equations is already complex, one should realize it would become even more intricate if movement was performed in the vertical plane, since this time a gravitational torque would also contribute to the shoulder and elbow net torque, the latter one being again a non-linear function of the arm orientation.

Ultimately the question emerges, as to how the brain deals with this complexity arising from mechanical laws if it explicitly controls muscle force? The assumption made by the force control hypothesis is that the brain has some ways of evaluating the respective contribution of each force prior to movement initiation. More specifically, it is assumed that the brain possesses some internal representations of the body mechanics interacting with the environment. Those representations, also called internal models, would allow the brain to perform inverse dynamic computations so as to determine the adequate muscle forces needed to achieve the desired movement (Kawato, 1999). More details will be provided on internal models in the next section.

Even though the concept of internal representations may offer a solution to muscle force computation, certain problems persist when controlling explicitly muscle force (Feldman & Levin, 1995; Ostry & Feldman, 2003). One of the concerns raised by Feldman and colleagues relates to the fact that switching from one muscle force to another does not always guarantee that we can make the intended movement. Let's imagine that you want to flex or extend your elbow, and that the initial position of your forearm is such that it is parallel to the floor (see Figure 5). In this position the gravitational torque is maximal and the muscle torque acting at the elbow must compensate for the gravitational torque. As we

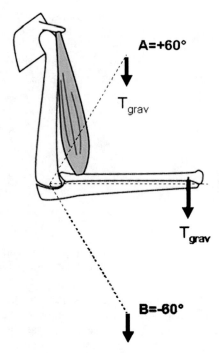

Fig. 5 Ambiguity of force control models when switching between postures. In posture A and B, the gravitational torque (T_{grav}) is the same; as a result the muscle torque needed to stabilize the arm in A and B are identical.

move away from this position, either upwards or downwards, the gravitational torque decreases. In fact, when the elbow stabilizes at 60° of flexion or extension, both postures can be maintained with 50% of the original muscle torque. Now how does the brain discriminate between these two postures if they both require the same muscle torque? And how can we switch unambiguously from the initial position to position A or B if they both require the same final muscle torque?

One advantage of the equilibrium point approach is that posture A and B are determined by distinct R commands, thereby avoiding any ambiguity between postures and central commands. In contrast, with the force model approach those postures cannot be distinguished since they are based on similar motor commands. Note that this problem (setting postures) would be even more prominent in the absence of external forces. Indeed, in the absence of gravity (i.e. horizontal plane), all flexed and extended postures of the elbow can be maintained with the same (null) muscle torque (Ostry & Feldman, 2003). One option to circumvent this problem is that the brain may not only be concerned with the final muscle torque, but also with the successive muscle torques that leads to the final muscle torque. Taking into account the action of inertial load acting on the elbow, patterns of muscle torques can be separated for movement A and B. If the hand goes up (elbow flexion) by 60°, muscle torque will first increase and then decrease. In contrast, if the hand goes down (elbow extension) by 60° muscle torque will first decrease and then increase. Altogether prescribing the time course of muscle torque may offer a solution to this problem. Obviously, as movement speed is decreased, the influence of the inertial load becomes smaller and smaller, and the similarity between the two patterns of muscle torque increases.

Another weakness of the force control hypothesis is its inherent instability with respect to noise in motor signals. In the earlier example, serious drift in position can occur if the muscle torque at the elbow fluctuates. In contrast, with positional control, even if R and C commands fluctuate, movement drift will be much more restricted. At last, it is unclear how the force control hypothesis deals with muscle cocontraction (Ostry & Feldman, 2003), since what matters first is the difference between the agonist and the antagonist muscle torques (i.e. net muscle torque), while their respective values hardly matter. Still, because muscle cocontraction is known to change as a function of learning, accuracy constraints, or expectation about perturbations, it is often considered as a key variable for movement control (Gribble et al, 2003; Darrainy & Ostry, 2008). Further developments of the force control hypothesis suggest that inverse dynamics models could be assisted by an impedance controller dealing specifically with the issue of muscle cocontraction and related limb stiffness (Franklin et al, 2003).

4 Mapping between Movements and Control Signals

In the previous sections we have exposed three different theoretical approaches that account for rhythmic movements. Let us now compare more explicitly those approaches in term of the underlying control signals. We would like to particularly insist on the nature of these control signals as well as the complexity of their patterns in relation to movement kinematics (see Figure 6). In the dynamical systems approach, a rhythmic movement is sustained without the need of time-varying input signals. Coming back to equation 2 and 3, as long as the parameters are held constant, the limb is supposed to keep oscillating at the same frequency and amplitude. In contrast, the equilibrium point approach and the computational approach both require time-varying input signals. If the underlying pattern of control signals can remain relatively simple in the equilibrium point approach (Gribble et al., 1998; Latash, 1993), it can become much more complex in the computational approach. Indeed, after removing the influence of external torques (such as gravity, or interaction torques) and other non-linearities (such as muscle moment arm), the net muscle force can exhibit a very complex pattern.

To better illustrate the influence of time-varying input signals versus constant input signals, we propose to address the effect of a subtle mechanical perturbation that would temporarily impede an oscillatory movement (i.e. then leading to a phase lag with respect to initial kinematics). Assuming that humans can maintain input signals despite the perturbation, we can make distinct predictions about how the oscillatory movement is restored after the perturbation. In the case of non-time varying input signals, as provided by equation 2 and 3, we predict that the phase lag should persist well after the perturbation (see the example in upper row of Figure 7). In contrast, with time-varying input signals, the longer cycle will be

Control signals for a rhythmical movement

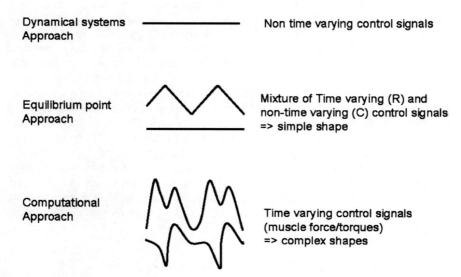

Fig. 6 Control signals used for a rhythmical movement in various approaches in motor control.

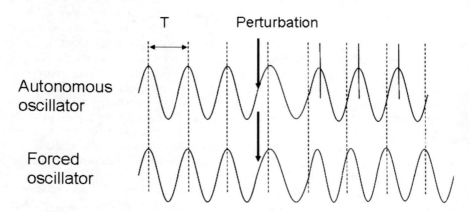

Fig. 7 Effect of a subtle perturbation that creates a phase lag in an oscillatory movement. The interval between two consecutive vertical lines corresponds to the period prior the perturbation. In the upper row, the phase lag is maintained after the perturbation (phase resetting), whereas in the lower row, the phase lag is compensated by faster cycles (no phase resetting).

followed by shorter cycles so that the initial phase is progressively restored (see the example in lower row of Fig 7). In fact, the time varying signals act as a kind of "internal clock" that makes its phase more robust to external perturbation. In physics, the first example corresponds to an autonomous oscillator, whereas the second example corresponds to a forced oscillator. To date, experimental data provide evidences of phase resetting with transcranial magnetic stimulation (Wagener & Colebatch, 1996; Latash et al, 2003), but it is very unlikely that input signals were preserved by this procedure.

In Figure 8, we compare changes in controlled parameter(s) underlying a rhythmic movement across the three theoretical approaches addressed in this chapter. The aim of this figure is to emphasize how the hypothetical controlled variable can be more or less closely related to the desired output (i.e. movement kinematics) in the different theoretical approaches. At the top of Figure 8, the dynamical systems approach proposes that the transformation from intention to movement is mediated by changes in abstract variables such as global stiffness or viscosity. Such changes may be followed by other changes in EMG, torques, reflexes, etc, but the latter are not explicitly controlled by the brain. At the other extremity (see bottom of Figure 8), the force control hypothesis suggests that the brain is explicitly concerned with fine details of the movement such as EMG and muscle torques that can be explicitly measured. The equilibrium point approach provides an intermediate position in which changes in EMG and muscle torques emerge from changes in reflex thresholds which are considered the key variables.

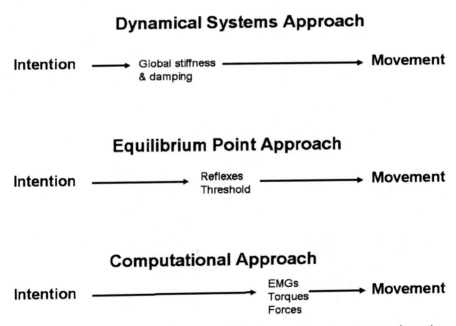

Fig. 8 Relation between intention, controlled parameter(s), and movement, in various approaches in motor control.

There is no doubt that the processes underlying the transformation of an intention into an action are complex. However, depending on the theoretical approach, this complexity is accounted for differently by the properties of the control signal, and the controller (providing the adequate control signal). For the dynamical systems approach, the controller seems to provide relatively simple patterns of control signals, but the causal relationship between the underlying control signals and their effects on movement related parameters (such as muscle torques and EMGs) is less trivial. In contrast, with the force control hypothesis, the relation between movement and changes in control parameters is much straightforward (i.e. explained by the laws of mechanics), but the issue of how the brain specifies the appropriate patterns of input signals becomes less trivial. Altogether, none of these two theoretical frameworks really offers a simple solution to movement control. In the first case, movement complexity is rather accounted by the abstract properties of the control signals, whereas in the second case it is essentially taken care by the smart design of the controller. In contrast to these approaches, the equilibrium point approach offers an intermediate opinion under which movement complexity is accounted (more) equally by the properties of the control signals and the controller.

Despite many differences across theoretical accounts, I would like to point out that, no matter the exact nature of the strings being pulled by the brain to produce a movement, the brain must know something about the relationship between those strings and their behavioral consequences when they are pulled. As a result, the issue of which strings should be pulled, and eventually how hard it should be pulled, is a question that can be formulated for each of the three approaches. Interestingly, while this question has been addressed frequently by the computational approach, it has been addressed less in the context of the dynamical system and equilibrium point approaches.

Another important related issue is that the mapping between motor commands and their effect on our body can change over time. The properties of our body change, and do so over different time scales (Newell et al, 2001). Indeed, during development, although our body grows slowly, we are confronted with geometric and inertial changes in our segments, as well as changes in our muscle force capabilities (Shadmehr & Mussa-Ivaldi, 1994). In adults, body properties can also change due to physical training or weight loss, and sometimes quite rapidly due to phenomena such as muscle fatigue. Note that humans also like using tools, carrying objects, wearing shoes, which turns out to affect the inertia of their segments, and therefore the mapping between motor commands and movement (Shadmehr & Mussa-Ivaldi, 1994). As a result, it may not be surprising that teenagers become temporarily clumsy when their body properties change relatively rapidly, or that during pregnancy women exhibit larger postural instabilities (Jang et al, 2008). Now changes in the external environment can also alter the way our body responds to motor commands. Lets imagine that you walk or ride a bike, and that suddenly strong gusts of wind start to blow from the left side of the road. It is very unlikely that you will be able to maintain your speed and direction if the motor commands sent to your arm and legs do not take into account this new force. The same reasoning applies when we want to perform the same movement in the air and under water, with or against gravity. Therefore, it should be made clear that there is no such thing as a unique mapping between motor commands and a resulting movement. Depending on the conditions (e.g. body properties and external forces), one motor command can give rise to two different movements, and conversely the same movement can originate from different motor commands. This conclusion fits well with the pioneering work of Nicholai Bernstein (1967): *"the relationship between movements and the innervational impulses which evoke them is extremely complex and is, moreover, by no means univocal"* (p.15).

5 Internal Models versus Look Up Tables

In the previous section we have discussed significant differences across motor control approaches. However, we have also pointed that all those approaches are based upon the existence of parameters whose value must be changed to initiate, maintain, and/or terminate an action. Let us take the example of a simple goal directed movement like that of our hand when we want to grasp an object. Each time we initiate such an action, our movement tend to be straight (Morasso, 1981), such that the initial direction of our hand is oriented toward the object (Sarlegna et al, 2004; see also Figure 10A). Because during the very first part of the movement (<50 ms), the brain has no access yet to sensory feedback resulting from the movement, this part of the movement can only be controlled in a feedforward manner. On top of demonstrating our ability to localize objects, this observation demonstrates that the brain possesses some knowledge about the relationship between (changes in) input signals and their effect on our body. Sometimes this knowledge can be partly of genetic origin (e.g. walking, swallowing, breathing…), but in many cases prior experience is necessary to acquire this knowledge. The question we would like to ask now is how this knowledge is stored in our brain? As proposed by Shadmehr and Mussa-Ivaldi (1994), there are two main options. A first option is to keep trace of behavioral (sensory) consequences induced by each parameter change that we have previously experienced, and to build a lookup table. A second option is using prior experience to create some kind of algorithm/function/internal model that allows for the computation of the relationship between our motor commands and their behavioral consequences.

To better illustrate the concepts of internal model and lookup table, let us take the following mathematical example (see Figure 9). When being asked for the product of 5 times 5, anyone (I suspect) can respond 25 very quickly and without the need of doing some mental calculus. In fact, we just learnt the result of this operation (as well as many other ones) by heart when we went to elementary school. On the other hand, when being now asked what the product of 12 by 27 is, you will certainly take a longer time to respond, and start doing some kind of computation because the results of this operation is not stored in your memory. Still, most of us are able to provide the correct response (i.e. 324). This is possible because what is stored in our brain is a method (or algorithm) that allows us to compute the product of 2 digit numbers, or more. Ultimately we can consider that

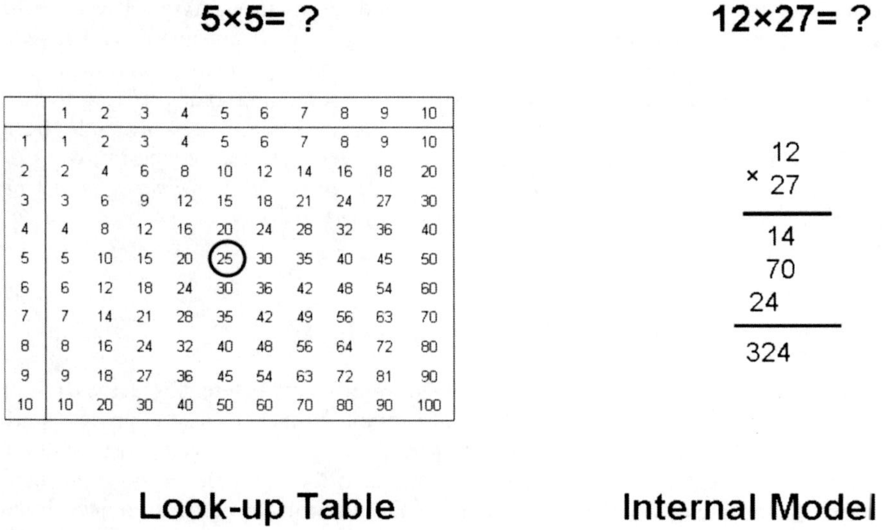

5×5= ?

	1	2	3	4	5	6	7	8	9	10
1	1	2	3	4	5	6	7	8	9	10
2	2	4	6	8	10	12	14	16	18	20
3	3	6	9	12	15	18	21	24	27	30
4	4	8	12	16	20	24	28	32	36	40
5	5	10	15	20	(25)	30	35	40	45	50
6	6	12	18	24	30	36	42	48	54	60
7	7	14	21	28	35	42	49	56	63	70
8	8	16	24	32	40	48	56	64	72	80
9	9	18	27	36	45	54	63	72	81	90
10	10	20	30	40	50	60	70	80	90	100

Look-up Table

12×27= ?

$$\begin{array}{r} 12 \\ \times\ 27 \\ \hline 14 \\ 70 \\ 24 \\ \hline 324 \end{array}$$

Internal Model

Fig. 9 Look up table versus internal model approach when dealing with multiplication.

what we use in the first case is a look up table, whereas it is more like an internal model in the second case. Note that without an internal model, we would be forced to store the results of all the possible pairs in our memory. The neat advantage of having an internal model is that we can provide correct answers to questions you have never been asked before. On the other hand, if someone was taught an inadequate method to compute 12 by 27, we expect this person to make errors for many other operations. This is where the lookup table may become more advantageous. Indeed, errors (can) remain local in a lookup table. We can easily imagine that someone being taught that 5 times 5 equals 26 can still correctly answer to what is the product of 5 times 4 or 5 times 6. These local versus global effects have been exploited to test whether the knowledge about our body and its environment is stored in lookup tables or in internal models (Shadmehr and Mussa-Ivaldi, 1994).

The need to update the knowledge between changes in parameters and their consequences on the body is particularly obvious in the context of changes in external forces. By means of a robot arm, researchers can apply various external force fields ranging from elastic, viscous, to even more complex ones (Shadmehr and Mussa-Ivaldi, 1994; Shadmehr and Brashers-Krug, 1997; Malfait et al, 2002, 2005). When subjects are confronted with unusual force fields, they typically experience difficulties in making straight movements and reaching targets (see Figure 10B). This difficulty arises from the fact that the 'natural' mapping between motor commands and their behavioral consequences has been modified through the action of the force field. Nevertheless, after a period of training, subjects can ultimately restore movement accuracy (see Figure 10C). A crucial

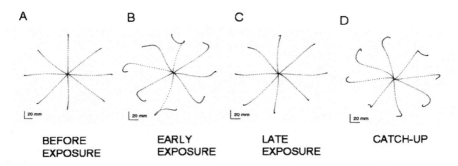

Fig. 10 Hand trajectories during exposure to a force field. A. No force field is generated by the robot arm. B. Early exposure to force field. C. Late exposure to force field. D. Trials in which the force field is unexpectedly removed. Movements originate at the center. Dots are 10 msec apart (adapted with permission from Shadmehr & Brashers-Krug, 1997).

feature of this learning phase is that if the force field generated by the robot arm is unexpectedly removed (so that initial conditions prior to the perturbation are restored), subjects start making again inaccurate movements, although now they deviate in the opposite direction (see Figure 10D). The presence of aftereffects is consistent with the view that some adaptive processes occurred during the task. In other words, aftereffects demonstrate that the knowledge between motor commands and their consequences has been updated during the learning phase.

To disentangle the update of an internal model and the update of lookup table, the trick used by experimenters is to test the generality of the knowledge is that is acquired during force field exposure (Shadmehr, 2004; Malfait et al, 2002, 2005). For instance, during the learning phase subjects are asked to make reaching movements toward a fixed number of targets, then when accuracy in performance is restored, subjects are tested with a new set of targets. The rationale is that if subjects are still able to make accurate movements, even though they have no previous experience with these new targets, this means that the knowledge acquired by the subjects is not supported by a lookup table. The update (or building) of an internal model remains the only option to account for the generalization of learning. The fact that Conditt and colleagues (1997) observed that subjects were able to accurately draw circles after adaptation to a viscous field while making reaching movements supports the notion of an internal model. Using a slightly different methodology, Shadmehr and Mussa Ivaldi (1994) also support this view when reporting that subjects exhibited aftereffects in regions of the workspace that were not experienced during adaptation.

Since this seminal study, many experiments have tried to circumscribe the generality of the knowledge acquired by motor learning. In most cases, results show that learning is rarely restricted to the exact conditions in which it was obtained, but at the same time they also point out some limitations in generalization. For instance, if one subject learned to compensate a force field with the right arm, he or she does not necessarily know how to make accurate movements with the left arm (Malfait & Ostry, 2004). In fact, this transfer between limbs appears relatively context dependent (Malfait & Ostry, 2004; Nozaki et al, 2006). Obviously, one possibility is that certain internal models are limb specific, while others are not (Morton et al, 2001).

6 Conclusion

Our initial question was do we need internal models for movement control? In the previous sections, I have reviewed and compared three approaches to motor control. Although opinions can diverge substantially about the nature of the variables that are effectively manipulated by the brain during a voluntary movement, at one point or another, all three approaches need to cope with the issue of parameter selection. They also need to cope with the fact that the mapping between parameter changes and their effects on the body is complex, redundant, and often need to be updated so as to preserve movement accuracy. Whether this mapping is organized in lookup tables or internal models is a question that will need further investigations. In any case, I hope that the information provided in this chapter made clear that the concept of internal model addresses a fundamental issue for anyone being interested in motor control. At last, although the concept of internal model is often restricted to the force control approach, I hope this chapter can extend its relevance to other approaches as well.

References

Abend, W., Bizzi, E., Morasso, P.: Human arm trajectory formation. Brain 105(2), 331–348 (1982)
Almeida, G.L., Hong, D.A., Corcos, D., Gottlieb, G.L.: Organizing principles for voluntary movement: extending single-joint rules. Journal of Neurophysiology 74, 1374–1381 (1995)
Ashe, J.: Force and the motor cortex. Behavioural Brain Research 87, 255–269 (1997)
Beek, P.J., Rikkert, W.E.I., van Wieringen, P.C.W.: Limit cycle properties of rhythmic forearm movements. Journal of Experimental Psychology: Human Perception & Performance 22, 1077–1093 (1996)
Bennett, D.J., Hollerbach, J.M., Xu, Y., Hunter, I.W.: Time-varying stiffness of human elbow joint during cyclic voluntary movement. Experimental Brain Research 88, 433–442 (1992)
Bernstein, N.A.: The co-ordination and regulation of movements. Pergamon Press, Oxford (1967)
Conditt, M.A., Gandolfo, F., Mussa-Ivaldi, F.A.: The motor system does not learn the dynamics of the arm by rote memorization of past experience. Journal of Neurophysiology 78, 554–560 (1997)
Dai, T.H., Liu, J.Z., Sahgal, V., Brown, R.W., Yue, G.H.: Relationship between muscle output and functional MRI-measured brain activation. Experimental Brain Research 140, 290–300 (2001)
Feldman, A.G.: Superposition of motor programs. I. Rhythmic forearm movements in man. Neuroscience 5, 81–90 (1980)

Feldman, A.G.: Once more on the equilibrium-point hypothesis (lambda model) for motor control. Journal of Motor Behavior 18, 17–54 (1986)

Feldman, A.G., Levin, M.F.: The origin and use of positional frames of reference in motor control. Behavioural and Brain Sciences 18, 723–806 (1995)

Flanagan, J.R., Tresilian, J.R., Wing, A.M.: Grip force adjustments during rapid hand movements suggest that detailed movement kinematics are predicted. Behavioural and Brain Sciences 18, 753–754 (1995)

Franklin, D.W., Osu, R., Burdet, E., Kawato, M., Milner, T.E.: Adaptation to stable and unstable dynamics achieved by combined impedance control and inverse dynamics model. Journal of Neurophysiology 90, 3270–3282 (2003)

Georgopoulos, A.P., Ashe, J., Smyrnis, N., Taira, M.: The motor cortex and the coding of force. Science 256, 1692–1695 (1992)

Gottlieb, G.L., Corcos, D.M., Agarwal, G.C.: Organizing principles for single-joint movements. I. A speed-insensitive strategy. Journal of Neurophysiology 62, 342–357 (1989)

Gottlieb, G.L.: Shifting frames but the same old point of view. Behavioural and Brain Sciences 18, 758 (1995)

Gottlieb, G.L.: Rejecting the equilibrium-point hypothesis. Motor Control 2, 10–12 (1998a)

Gottlieb, G.L.: Muscle activation patterns during two types of voluntary single-joint movement. Journal of Neurophysiology 80, 1860–1867 (1998b)

Gottlieb, G.L., Agarwal, G.C.: Dynamic relationship between isometric muscle tension and the electromyogram in man. Journal of Applied Physiology 30, 345–351 (1971)

Gribble, P.L., Ostry, D.J., Sanguineti, V., Laboissiere, R.: Are complex control signals required for human arm movements? Journal of Neurophysiology 79, 1409–1424 (1998)

Gribble, P.L., Ostry, D.J.: Compensation for loads during arm movements using equilibrium-point control. Experimental Brain Research 135, 474–482 (2000)

Gribble, P.L., Mullin, L.I., Cothros, N., Mattar, A.: Role of cocontraction in arm movement accuracy. Journal of Neurophysiology 89, 2396–2405 (2003)

Hill, A.V.: The heat of shortening and the dynamic constants of muscle. Proceedings of the Royal Society of London B 126, 136–195 (1938)

Humphrey, D.R., Reed, D.J.: Separate cortical systems for control of joint movement and joint stiffness: reciprocal activation and coactivation of antagonist muscles. Advances in Neurology 39, 347–372 (1983)

Kawato, M.: Internal models for motor control and trajectory planning. Current Opinion in Neurobiology 9, 718–727 (1999)

Jang, J., Hsiao, K.T., Hsiao-Wecksler, E.T.: Balance (perceived and actual) and preferred stance width during pregnancy. Clinical Biomechanics 23, 468–476 (2008)

Kay, B.A., Kelso, J.A., Saltzman, E.L., Schöner, G.: Space-time behavior of single and bimanual rhythmical movements: data and limit cycle model. Journal of Experimental Psychology: Human Perception & Performance 13, 178–192 (1987)

Latash, M.L.: Virtual trajectories, joint stiffness, and changes in the limb natural frequency during single-joint oscillatory movements. Neuroscience 49, 209–220 (1992)

Latash, M.L.: Control of Human movement. Human Kinetics (1993)

Latash, M.L., Zatsiorsky, V.M.: Joint stiffness: Myth or reality. Human Movement Science 12, 653–692 (1993)

Latash, M.L., Danion, F., Bonnard, M.: Effects of transcranial magnetic stimulation on muscle activation patterns and joint kinematics within a two-joint motor synergy. Brain Research 961, 229–242 (2003)

Lippold, O.C.J.: The relation between integrated action potential in human muscle and its isometric tension. Journal of Physiology (London) 117, 492–499 (1952)

Malfait, N., Shiller, D.M., Ostry, D.J.: Transfer of motor learning across arm configurations. Journal of Neuroscience 22, 9656–9660 (2002)

Malfait, N., Gribble, P.L., Ostry, D.J.: Generalization of motor learning based on multiple field exposures and local adaptation. Journal of Neurophysiology 93, 3327–3338 (2005)

Malfait, N., Ostry, D.J.: Is interlimb transfer of force-field adaptation a cognitive response to the sudden introduction of load? Journal of Neuroscience 24, 8084–8089 (2004)

Morasso, P.: Spatial control of arm movements. Experimental Brain Research 42, 223–227 (1981)

Morton, S.M., Lang, C.E., Bastian, A.J.: Inter- and intra-limb generalization of adaptation during catching. Experimental Brain Research 141, 438–445 (2001)

Mottet, D., Bootsma, R.J.: The dynamics of goal-directed rhythmical aiming. Biological Cybernetics 80, 235–245 (1999)

Newell, K.M., Liu, Y.T., Mayer-Kress, G.: Time scales in motor learning and development. Psychological Review 108, 57–82 (2001)

Nozaki, D., Kurtzer, I., Scott, S.H.: Limited transfer of learning between unimanual and bimanual skills within the same limb. Nature Neuroscience 9, 1364–1366 (2006)

Olney, S.J., Winter, D.A.: Prediction of knee and ankle moments of force in walking from EMG and kinematic data. Journal of Biomechanics 18, 9–20 (1985)

Ostry, D.J., Feldman, A.G.: A critical evaluation of the force control hypothesis in motor control. Experimental Brain Research 153, 275–288 (2003)

Pigeon, P., Yahia, L., Feldman, A.: Moment arms and lengths of human upper limb muscles as functions of joint angles. Journal of Biomechanics 29, 1365–1370 (1996)

Rosenbaum, D.A.: Is Dynamical systems modelling just curve fitting? Motor Control 2, 101–104 (1998)

Sarlegna, F., Blouin, J., Vercher, J.L., Bresciani, J.P., Bourdin, C., Gauthier, G.M.: Online control of the direction of rapid reaching movements. Experimental Brain Research 157, 468–471 (2004)

Shadmehr, R.: Generalization as a behavioral window to the neural mechanisms of learning internal models. Human Movement Science 23, 543–568 (2004)

Shadmehr, R., Mussa-Ivaldi, F.A.: Adaptive representation of dynamics during learning of a motor task. Journal of Neuroscience 14, 3208–3224 (1994)

Shadmehr, R., Brashers-Krug, T.: Functional stages in the formation of human long-term motor memory. Journal of Neuroscience 17, 409–419 (1997)

Sainburg, R.L., Ghez, C., Kalakanis, D.: Intersegmental dynamics are controlled by sequential anticipatory, error correction, and postural mechanisms. Journal of Neurophysiology 81, 1045–1056 (1999)

Solomonow, M., Baratta, R., Zhou, B.H., Shoji, H., D'Ambrosia, R.: Historical update and new developments on the EMG-force relationships of skeletal muscles. Orthopaedics 9, 1541–1543 (1986)

Thelen, D.G., Schultz, A.B., Fassois, S.D., Ashton-Miller, J.A.: Identification of dynamic myoelectric signal-to-force models during isometric lumbar muscle contractions. Journal of Biomechanics 27, 907–919 (1994)

Wagener, D.S., Colebatch, J.G.: Voluntary rhythmical movement is reset by stimulating the motor cortex. Experimental Brain Research 111, 113–120 (1996)

Nonlinear Dynamics in Speech Perception

Betty Tuller, Noël Nguyen, Leonardo Lancia, and Gautam K. Vallabha

Abstract. The history of research on speech perception and speech production is replete with examples of nonlinearities between articulation and acoustics, and between acoustics and perception. These nonlinearities are useful for communication. They allow 1) adequate production of speech sounds and words despite people having different vocal tracts with different resonance capabilities, and 2) adequate word recognition despite variation in the acoustic signal across speakers, emphasis, background noise, etc. Yet context and the listener's expectancies often strongly influence what is perceived; perception is dynamic, influenced by multiple factors that change slowly or quickly as speech goes on. In this chapter we present a selected history of demonstrations of nonlinearities in speech and attempt to exploit the nonlinearities in order to uncover the dynamics of both perception and production of speech.

1 Introduction

Speech perception depends largely on information in an acoustic stream that is inherently dynamical in that it changes constantly and fades rapidly. Yet many of the psychoacoustic studies that formed the basis for the exploration of speech acoustics historically used pure or complex tones with little temporal change. In this chapter, we present selected examples of nonlinearities in speech production and perception, especially with regard to context sensitivity, stability, and flexibility. Next, we present work that exploits these nonlinearities in order to explore the dynamics. Throughout the chapter, we provide links to a web site with demonstrations of many of the perceptual phenomena (file:///C:/data/ Springer%20book/tuller/tuller_et-al/index.html).

Betty Tuller
Florida Atlantic University, Boca Raton FL and
National Science Foundation, Arlington VA

Noël Nguyen
Laboratoire Parole et Langage, Aix-Marseille Université
CNRS 5, Avenue Pasteur, 13100 Aix-en-Provence, France

Leonardo Lancia
Max Planck Institute for Evolutionary Anthropology Leipzig, Germany

Gautam K. Vallabha
Florida Atlantic University, Boca Raton, FL; currently with
The MathWorks, Natick, MA USA

R. Huys and V.K. Jirsa (Eds.): Nonlinear Dynamics in Human Behavior, SCI 328, pp. 135–150.
springerlink.com © Springer-Verlag Berlin Heidelberg 2010

2 Nonlinearity

Psychoacoustics refers to the psychological (subjective) correlates of the physical parameters of acoustics. Much of the work on psychoacoustics has been based on non-speech stimuli, such as pure or complex tones, that vary little or not at all over time (see B.C. J. Moore, An introduction to the psychology of hearing, for a good introduction to psychoacoustic research). In general, the relationship between a physical aspect of the acoustic signal and its subjective correlate is logarithmic, following the patterns described by E.H. Weber and G.T. Fechner in the 19th century (see also S.S. Stevens, 1946). The logarithmic relationship means that as the physical stimulus increases in a geometric progression, the subjective perception changes in an arithmetic progression (see Gescheider, 1997 for review). In other words, each additional step change in the physical stimulus corresponds to a relatively smaller perceptual change. For example, the frequency of corresponding notes in adjacent octaves differs by a factor of two. This means that although the perceived pitch relationship between middle C and C one octave higher is the same as the pitch difference between D above middle C and D in the next higher octave, the absolute frequency difference is smaller between the successive C notes (~ 261.63 Hz and 523.25 Hz, respectively) than between the Ds (~ 293.66 Hz and 587.33 Hz, respectively). Another well-known logarithmic scale describing a psychoacoustic relationship is the decibel scale, which captures the relationship between sound intensity and loudness.

Another way of looking at these psychoacoustic relationships is by asking the question, "For a given value of an acoustic parameter (e.g., frequency, intensity, or duration), how small of a difference in that parameter can the human auditory system detect?" Intriguingly, the answer depends on the type of acoustic signal. In general, the size of the minimal acoustic change that can be detected (the so-called "just noticeable difference," or JND), is larger for speech or speech-like stimuli than for pure tones or, in some cases, noise. JNDs for loudness of pure tones or wideband noise at amplitudes in the speech range are about 0.3-1.0 dB. The JND for loudness of a vowel's second formant (a formant is a frequency band in the vowel that is of relatively high energy) is much larger, namely about 3dB (K.N. Stevens, 1998).

The larger JNDs for speech may contribute to the remarkable stability of speech perception across acoustic variation due to context, speaker, speaking rate, and so on. But there is an important underlying principle implied by the larger JNDs, namely, a strong nonlinearity in the relationship between speech acoustics and perception. This nonlinearity is exemplified by the extreme difficulty native speakers have in discriminating between acoustically different speech stimuli that are categorized as the same linguistic segment (see Repp, 1984, for a review). We should note that discrimination is not difficult to the same extent for all types of speech stimuli, e.g., vowels are more easily discriminated than are stop consonants. One of the early demonstrations of nonlinear perceptual pattern formation in speech was provided by Liberman, Harris, Hoffman, and Griffith (1957) who presented listeners with synthetic syllables consisting of a consonant followed by a synthesized approximation to the vowel in "gate". The vowel was

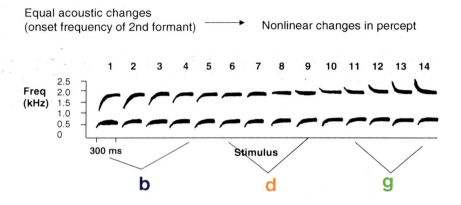

Adapted from Liberman, Harris, Hoffman, and Griffith (1957)
Journal of Experimental Psychology 54, 358-368

Fig. 1 When the onset frequency of the second formant changes in equal sized steps, perception of the initial consonant's identity changes abruptly.

not very life-like in that the energy resonances (the formant trajectories) did not change as the vowel progressed, which they would do in a naturally-produced vowel. Liberman et al. varied the onset frequency of the second formant (F2) in equal steps and found that listeners' perceptions did not change equally with each acoustic step change. Instead, several syllables with different F2 were perceived as "bay." After the F2 frequency reached a critical value, listeners perceived the syllable as "day" and, after a few more step changes in F2 onset frequency, they perceived "gay" (Figure 1). That is, a sequence of equal acoustic changes produced nonlinear shifts in perception. Liberman et al. confirmed that the results did not stem from listeners having no label for intermediate speech sounds; when asked simply to discriminate between stimuli, listeners discriminated worst when they had identified two stimuli as the same speech sound. Another way to think of this is that the JNDs were not equal across the stimulus set but were smallest at category boundaries.

This diminished ability to discriminate between speech stimuli that are categorized as the same speech sound is termed categorical perception and has been replicated many times in the intervening years with different stimulus continua and in different contexts. Outside the laboratory as well, a wide range of acoustic instances are categorized identically, such as when the same word is produced by different speakers. The boundaries between categories are flexible and adjust with factors such as individual speaker differences, different listeners, phonetic context, speaking rate, and linguistic experience (Repp & Liberman,

1987). Nevertheless, the shifts in the perceived identity of a sound tend to be abrupt rather than continuous. An easy way to get a sense of this phenomenon is to make the sound "s" (as in the word "sip"). While continuing to blow air through your mouth so you still hear the sound, slowly move your tongue back along your palate. You will probably still perceive that you are making the "s" sound even though your tongue has moved back a little. At some point, however, you will perceive an abrupt change to the "sh" sound (as in "ship"). Moving the tongue even further back will have little effect on the identity of the "sh" sound (although the quality changes and there is, of course, a limit). An acoustic example may be found as demonstration 1 at the website link.

3 Context Sensitivity

The perceptual dynamics of speech are not simply nonlinear but also highly context dependent. Different acoustic patterns can be perceived as the same phoneme when they appear in their acoustically appropriate context. Similarly, different articulatory patterns are used to produce the same phoneme in different contexts[1]. In 1966, Öhman studied the articulatory movements that occur in vowel-consonant-vowel sequences. He observed that the tongue position for a /d/ was quite different when it was surrounded by different vowels. Figure 2, adapted from Öhman (1966), shows contour tracings from x-ray motion pictures for the consonant /d/ surrounded by the vowel /y/ (as in the French "du") top graph), /a/ (middle), and /u/ (bottom, as in the English word "do"). The x-ray is at the point when the tongue is occluding the vocal tract for production of the consonant closure. Note the very different tongue positions for /d/ closure in the three different vowel contexts. The consonant production is contextually sensitive to the vowel environment.

To summarize this brief introduction, humans categorize a variety of speech signals (arising from a variety of articulatory events) identically, which gives communication stability across context. But humans can also categorize the identical acoustic signal differently depending on its context, rate, etc. In a now classic demonstration, Ladefoged and Broadbent (1957) synthesized six versions of the sentence "Please say what this word is ." Four test words were also synthesized ("bet," "bit," "bat," and "but"). A trial consisted of one version of the sentence followed by one test word. Identification of the test word strongly depended on the formant structure of the introductory sentence. In Ladefoged and Broadbent's words, "...the linguistic information conveyed by a vowel sound does not depend on the absolute values of its formant frequencies, but on the relationship between the formant frequencies for that vowel and the formant frequencies of other vowels pronounced by that speaker." This suggests that listeners "normalize" their percepts to the overall characteristics of a given speaker, allowing great flexibility in the service of communication. An acoustic example may be found as demonstration 2 at the website.

---

[1] The ability to obtain a particular task result via different motor means, referred to as motor equivalence, is a well-known phenomenon in motor control.

4 Stability and Flexibility

First, a definition. A stable system (in the present context) is one that is robust across non-linguistic variation (see Fuchs, chapter one, this volume for a more general treatment). That is, parts of the signal can be changed or even missing without disrupting understanding. This quality is complementary to the flexibility exemplified by the Ladefoged and Broadbent work described above, as veridical perception needs to be both stable *and* flexible. Speech can undergo many kinds of distortion, in addition to the contextual modulations introduced earlier, and remain intelligible. An example of this was provided by Warren in 1970 and was dubbed "phonemic restoration." If a segment in connected speech is replaced by noise,

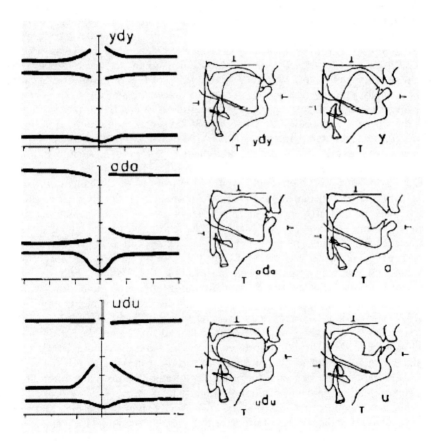

Fig. 2 Different articulatory patterns are used to produce the same phoneme in different contexts. Note the position of the tongue during the /d/ in the three vowel environments. Adapted from Öhman (1966)

listeners usually do not notice that it is missing, and cannot accurately report during which segment the noise occurred (Warren 1970, 1996). Phonemic restoration is strongly influenced by the semantic context. Warren and Warren (1976) performed a study in which they replaced a phoneme with a cough. In the following sentences, the * denotes the cough sound. Listeners heard a) The *eel was on the axle; b) The *eel was on the shoe; c) The *eel was on the orange; d) The *eel was on the table. They perceived a) The wheel was on the axle; b) The heel was on the shoe; c) The peel was on the orange; d) The meal was on the table. In other words, the semantic context was used to 'fill in' the phoneme that was (supposedly) obstructed by the cough. An example of this semantically guided phonemic restoration may be found as demonstration 3 on the website (from Bregman Audio CD demonstrations1:12).

Note that the effects of semantic and phonemic context can occur over an extended time. But the question of what information in the acoustic signal itself is allowing the global context sensitivity and stability is unclear. An intriguing line of research uses a form of speech synthesis called "sinewave synthesis" (Remez, Rubin, Pisoni, & Carrell, 1981) in which the formants in speech are replaced with pure tone whistles that follow the center frequency of the formants. Some listeners spontaneously perceive sinewave speech as continuous speech. Others perceive sinewave speech as nonspeech whistles and glides but can be primed so that they hear the stimulus as speech if they know what the sentence is supposed to be. This means that listeners can perceive speech without traditional speech cues, paving the way for a view of speech as a dynamic pattern of trajectories through articulatory-acoustic space (demonstration 4 on the website).

Our own work is motivated by the realization that changes in speech categorization that occur as the acoustic signal or psychological context is altered may be indicative of a pattern formation process with its own *perceptual dynamics* (multistability, loss of stability, switching, etc.). Perceptual dynamics characterizes the time-dependent behavior of the speech system in terms of its (nonlinear) dynamics; that is, equations of motion describing the temporal evolution of the perceptual process, especially the stability and change of perceptual forms. This approach is analogous to the theoretical framework of *coordination dynamics* used to understand goal-directed movements (e.g., Kelso, 1995). Coordination dynamics does not refer to biomechanics per se (e.g., masses and stiffnesses of moving segments) but rather characterizes the spatiotemporal patterns of coordinated actions produced by the nervous system which in turn produce dynamic (and in the present context, acoustic) trajectories. Empirical work becomes, in part, a search for principles of autonomous pattern formation in speech articulation, perception and brain function.

In the next sections, we describe our own work exploring the idea that speech dynamics involves the pattern of trajectories through articulatory-acoustic space as well as the time-dependent behavior of the perceptual system itself. This characterization helps account for the ability of the speech production/perception system to cope with the huge variety of phoneme realizations. We address this and other issues in the next sections describing empirical work on both vowels and consonants.

5 Dynamics of Vowel Perception and Imitation

The nature of vowel representation has long been a contentious topic. The perception of a vowel sound depends upon its phonemic status (e.g., a good versus bad exemplar of a category), phonetic quality (e.g., produced with a larger versus a smaller vocal tract constriction), and temporal dynamics (e.g., whether the acoustics vary or are constant over the course of the vowel). Part of the difficulty in assessing these different factors is that the methods typically used to evaluate vowel perception – categorization, discrimination, and rating – discretize a fundamentally continuous percept to varying degrees and this discretization may obscure some of the features of the representation. An alternative method is that of vowel imitation, in which the subject is presented with vowel-like targets of systematically varying phonetic quality, and asked to imitate them as closely as possible. The differences in formant location between the target and the imitation may then be examined for clues to the underlying vowel representations. This method may be especially apt for speech categories because there is a tight linkage between speech perception and production, manifested by accurate imitation of vowels at remarkably short response latencies (Porter and Lubker, 1980) and by spontaneous imitation of words during shadowing (Shockley, Sabadini, & Fowler, 2004). In addition, imitation has been posited to be central to language acquisition by children (Kuhl, 2000) and language evolution in general (Studdert-Kennedy, 2000). Here we explore whether for individuals, the dynamics of vowel perception and production are complementary.

Early studies that used imitation to study vowel representation (Chistovich et al, 1966; Kent, 1973; Repp & Williams, 1985) attempted to evaluate whether vowel representation was granular, and if so, whether the grain was phonemic or allophonic (an allophone is a variant of a phoneme; changing the allophone does not change the meaning of a word). In the earlier studies, synthesized vowel-like stimuli were uniformly spaced along a one-dimensional cut in a space whose dimensions were the first and second formant frequencies (energy concentrations in the spectra) and presented as targets for imitation. An example of a one-dimensional cut in this "F1 X F2 space" would range from /i/ as in "tea" to /u/ as in "to." The imitations gravitated toward certain regions of the F1 X F2 space, suggesting some granularity in the representations. However, these results were confounded by the synthetic nature of the targets, since a speaker cannot reproduce arbitrary formant patterns but only those that are physiologically possible for his or her vocal tract. Repp and Williams (1987) addressed this confound by presenting a speaker's own productions as targets for imitation. Remarkably, the response tendencies ("biases") persisted even with imitation of self-produced targets.

Vallabha and Tuller (2004) examined the biases by having subjects imitate synthetic vowel-like stimuli, systematically spaced in a two-dimensional grid over the F1 X F2 space (denoted as [V]). From the 100 imitations by each speaker, 45 were selected such that they were well distributed over the entire vowel space, and each speaker imitated his 45 self-produced targets. In addition, each speaker's productions of natural words were recorded so as to provide a "map" of the

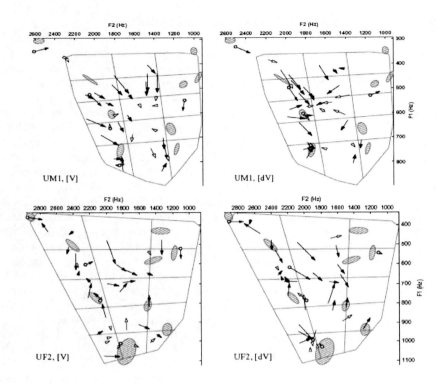

Fig. 3 The imitation behavior for one male (UM1) subject and one female subject (UF2) imitating their own productions of a vowel in isolation, or preceded by a /d/ (the [V] and [dV] conditions). The base of each arrow is a target and the tip is the mean of the 10 imitations of that target. Hatched regions are the 1-sd principal components for the /hVd/ productions. Solid arrowheads indicate statistically significant bias vectors (Hotelling's T^2 test, $p < .05$). Arrows with circles at their base indicate natural [V] and [dV] targets.

speaker's individual vowel prototypes. In a parallel experiment, Vallabha and Tuller preceded the vowel-like stimuli by the consonant /d/, forming /d/-vowel syllables ([dV]).

Figure 3 illustrates the bias vectors for one female and one male subject, both from South Florida. The vectors were calculated from the frequency values for F1 and F2 at the midpoint of each vowel in successive imitations. The significance of each vector was evaluated using Hotelling's T^2 test, which is a multivariate version of the z-test. Across subjects, the systematic biases produced were not attracted by the proximal vowel prototypes nor were they strongly driven by centralization. Note also that subjects with the same dialect did not necessarily have similar patterns of bias. This variety could not be attributed to noise, since most of the bias vectors were statistically significant, and the magnitudes of the biases were significantly larger than those predicted by a model of articulatory noise.

When we considered reproduction of acoustic trajectories, instead of "target values," we observed that subjects were remarkably accurate in reproducing the F1 and F2 trajectories in the more dynamic d-vowel syllables. Imitation of d-vowel syllables also showed less imitation variability than for the isolated [V] targets. Thus, although the early work on imitation (Chistovich et al, 1966; Kent, 1973; Repp & Williams, 1985, 1987) interpreted the "response preferences" as evidence for categorical representations, our results do not support such an interpretation. Nor do they support the notion that vowel-like sounds are represented with reference to immutable perceptual anchors or attractors. In fact, our results call into question the assumption that the vowel space is stable and organized around a few well-defined phonemic categories. It appears more likely that vowel-like sounds are represented with respect to the various linguistic and pragmatic contexts of vowel production. This perspective accounts naturally for the context-sensitivity and non-categoricality of vowel perception (e.g., Pisoni, 1973; Repp, Healy & Crowder, 1979; Repp & Crowder, 1990) and the importance of fine phonetic detail for perception (Hawkins, 2003). It can potentially also account for the perceptual relevance of dynamic trajectories through articulatory-acoustic space. A temporally varying sound can be more diagnostic of the linguistic and pragmatic context than a steady-state sound and therefore more valuable to a listener (see Muchisky & Bingham, 2002, for an analogous argument about "trajectory forms" in visual perception).

6 Dynamics of Consonant Perception: English and French

In our earlier discussion of categorical perception we described the nonlinearity between speech acoustics and consonant perception. But can we understand the nonlinear relationship between acoustics and categorization as a dynamic system? To reiterate, we are exploring whether dynamics can reconcile the highly context-dependent sensitivity of perception to the detailed acoustic structure of the speech input with the characterization of language as having a limited number of stable states. In a dynamical view, the stable states are not purely symbolic representations but themselves have a dynamic. The stable states are viewed as attractors, which allow the system to perform the discretization of perceptual space associated with abstract perceptual categories.

In our early experiments (Case, Tuller, Ding & Kelso, 1995; Tuller, Case, Ding & Kelso, 1994), stimuli on a *say-stay* continuum were presented to the listener in either a randomized order, or a sequential order. The individual tokens differed only in the duration of the short silent gap after the fricative noise of the /s/ sound. Stimuli with short (or absent) silent gaps were perceived as "say" whereas stimuli with long gaps were perceived as "stay." Monolingual speakers of American English heard sequential presentations of the entire set of stimuli, starting at one of the two endpoints (e.g., *say* with 0-ms silent gap), progressing to the other (*stay* with 76-ms silent gap), and then back again to the first one (*say*) with the gap duration of successive stimuli changing in 4-ms steps. Listeners were required to identify each stimulus as either "say" or "stay." The specific measures of interest were a comparison of where an individual listener's percept switched from one word to the other as the silent gap (the control parameter) increased or decreased, and the variability around the switch point. In sequential presentations, there are only three possible response patterns: *1)* a critical boundary, where the switch between the two percepts occurs at the same stimulus regardless of the direction of presentation across the continuum; *2)* hysteresis, defined as the tendency for the listener's response at one endpoint to persist across the ordered sequence of stimuli towards the other endpoint, and *3)* contrast, in which the listener quickly switches from the initial categorization. The results showed that critical boundary was much less frequent than hysteresis and contrast, which occurred equally often over the entire experiment, although hysteresis was far more frequent during the first half of the experiment. The observation of hysteresis and contrast implies bistability in perception: two percepts are possible for the same acoustic token. Moreover, the introduction of "noise" (for example, repetition of the same token) increased the likelihood of a perceptual switch only in the bistable region. These data provide strong support for the idea of speech perception as a process, characterized by a rich variety of dynamical properties. Readers are referred to Tuller et al. (1994) and Case et al. (1995) for further detail on these experiments and to the website for a demonstration (#5) of a sequential presentation of the continuum.

Nguyen, Lancia, Bergounioux, Wauquier-Gravelines & Tuller (2005) extended Tuller and colleagues' (1994) hypotheses and experimental paradigm to the categorization of speech sounds in French. They manipulated the same acoustic variable as did Tuller and colleagues, but this time the stimulus continuum ranged between the French words *cèpe* (a type of mushroom) and *steppe* (in physical geography, a steppe, that is, a plain without trees). Just as for the English listeners, French listeners perceived stimuli with short silent intervals after the /s/ noise as being followed directly by the vowel (*cèpe*) and perceived stimuli with longer silent gaps as having a /t/ after the /s/ (*steppe*). Native speakers of French listened to the stimuli and were asked to respond as quickly as possible after each stimulus whether they heard *cèpe* or *steppe* (demonstration #6 on website). Lancia, Nguyen & Tuller (2008) and Nguyen, Wauquier, & Tuller (2009) devised an index referred to as the Contrast-Hysteresis (CH) index to measure the amount that hysteresis or contrast contributed to each subject's responses to sequentially presented stimuli (Figure 4). This entailed locating the position on the continuum of the stimulus associated with the switch from one response to the other in the first part of the presentation and comparing it with the location of the switch in the second part of the presentation. The distance between these two points was then measured. Results showed that in French as in English, hysteresis and contrast prevailed over critical boundary (note that the CH index is positive in the right panel of Figure 4).

Nonlinear Dynamics in Speech Perception

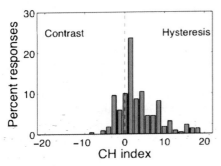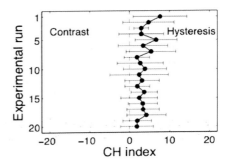

Fig. 4 The Contrast-Hysteresis (CH) index for perception of *cèpe* or *steppe* by French monolingual listeners during sequential presentation of stimuli. Left panel: distribution across all the presentations. Right panel: mean value and standard deviation for each of the 20 sequential presentations (from Nguyen, Wauquier, & Tuller, 2009).

These results confirm that the speech perception system can be modelled as a nonlinear dynamical system. A dynamical system is one that evolves over time such that its present state always depends in some rule-governed way on previous states. Differential equations or maps of essential variables offer a mathematical description of how a behavior's essential parameters change as time passes and contextual parameters change (spectral composition, rate of presentation/ production etc.). In such a system, phonological categories are equivalent to attractors (stable behaviors of the system) and switching between phonological categories means changes in the relative stability of the attractors. Thus, there exist ranges of acoustic parameter variation within which the perceptual form remains relatively stable (i.e. resistant to change as a function of parameter variation or noise). In other ranges, however, even small variations in the acoustic parameter can cause large (nonlinear) changes in categorization of the input and these perceptual changes are enhanced in the presence of noise. At these critical values, which are sensitive to context, history, linguistic factors, etc., the existing attractor(s) lose stability and the observed behaviors may change gradually or abruptly as other attractors dominate. Abrupt, or qualitative, changes are called phase transitions or bifurcations. Signature properties of dynamical systems (e.g. hysteresis) were observed for both French and English listeners in that the critical point for switching in any given trial depended on the direction of changing gap duration in the stimulus sequence and the initial percept. The switching between categories was modeled as the appearance and disappearance of attractive states in the underlying dynamical system such that changes in perceived category occur when the attractor corresponding to the initial category loses stability.

A simple dynamical model, developed to describe the results, has been useful as a tool for predicting patterns of categorization (see also Fuchs, chapter one, this volume). The model was designed to account for listeners' response patterns in the binary-choice speech categorization task. Here, we conceptualize articulatory/ perceptual motion towards a linguistic category under the influence of the forces supplied by the changing acoustics, task, intention, etc.

$$V(x) = kx - x^2/2 + x^4/4 \tag{1}$$

In equation (1), x represents the perceptual form (*say* vs. *stay*, or *cèpe* vs. *steppe*), k a control parameter and $V(x)$ a potential function which may have up to two stable perceptual forms, indicated by minima in the potential function, depending on the value of k. The control parameter k itself depends on the acoustic characteristics of the stimulus (the gap duration of the /s/ sound), and the combined effects of learning, linguistic experience and attentional factors, in a way described by the following equation:

$$k(\lambda) = k_0 + \lambda + \varepsilon/2 + \varepsilon\theta(n - n_c)(\lambda - \lambda_f) \tag{2}$$

where k_0 refers to the system's initial state, λ represents the acoustic parameter that is manipulated in the stimuli (in the present case, the duration of the silent gap between the /s/ noise and the vowel), ε is a parameter that characterizes the lumped effect of learning, linguistic experience and attention, $\varepsilon\theta(n - n_c)$ is the discrete form of the Heaviside step function, n is the number of perceived stimulus repetitions in a given run, n_c represents a critical number of accumulated repetitions (which we define as occurring at the turnaround point in a run, for lack of a more precise value), and λ_f denotes the value of λ at the other extreme from its initial value (so that, for example, if λ=0ms then, in our experiment, λ_f= 76ms). The value of λ and λ_f depend on the parameter range *and* the direction of parameter change.

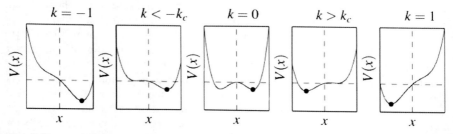

Fig. 5 Shape of the potential function $V(x)$ for five values of k. Adapted from Tuller et al. (1994).

For a given value of k, the system's state evolves in the x perceptual space to get trapped into a local minimum, or attractor, of $V(x)$. Each of the two possible responses in a binary categorization task corresponds to one attractor in the perceptual space. Figure 5 shows the shape of the potential function for five values of k between -1 and 1. The potential function has one minimum only for extreme values of k, which correspond to stimuli unambiguously associated with either of the two categories, and two minima in the middle range of k, where both categories are possible. As k increases in a monotonic fashion (from left to right in Figure 5), and in the vicinity of a critical value k_c, the system's state, represented by the filled circle in Figure 5, abruptly switches from the basin of attraction in which it was initially located, to the second basin that has gradually formed as the first one disappears. The stability of the attractor can be observed in the categorization data: fluctuation (variability) increases as the attractor becomes less stable, and decreases after the switch into the new, more stable attractor (a phase transition).

This model was developed to help understand the dynamics of changes in categorization with acoustic variation. However, the transition between categorizations has also shown to be an important entry point for understanding the acquisition of nonnative sounds (Tuller, Jantzen, & Jirsa, 2008). In fact, the model shows a number of general properties that, in our view, open the way towards a novel, hybrid view to speech perception that gets beyond the dichotomy traditionally established between symbolic and dynamic representations of speech perception (Nguyen, Wauquier, & Tuller, 2009). In recent years, this dichotomy has been couched in terms of abstractionist and exemplar-based approaches to speech perception. In an abstractionist approach, the speech signal is converted by listeners into a set of context-independent abstract phonological units. Variations in the acoustic instantiation of a given word, including cross-speaker differences are thought to be factored out at an early stage of processing (see Johnson, 2005 for review). In an abstractionist framework, there is a clear demarcation between the surface phonetic form of a word and the underlying phonological representation associated with that word.

In the exemplar approach, each acoustic instantiation of a given word (an exemplar) accumulates in memory as listeners' exposure increases, causing the boundaries between categories to be continuously pushed around in perceptual space. As a result, categories represented by a larger number of exemplars gradually come to prevail over ones less frequently encountered. Perceptual categories are thus taken to be time-dependent and to evolve continuously in the course of the conversational interactions in which speakers/listeners engage. This of course has major theoretical implications, as frequency of use is expected to shape the form of phonological representations.

Experimental evidence supports a role for both exemplars and abstract phonological categories in speech perception. Hybrid models (Hawkins, 2003, 2007; Luce & McLennan, 2005; McLennan & Luce, 2005; Pierrehumbert, 2006) are governed by the assumption that abstract phonological categories and the fine phonetic detail that differentiates among exemplars of a category combine with each other in the representations associated with words in memory. The dynamical model outlined here is a form of a hybrid model.On the one hand, speech perception is assumed to be a highly context-dependent process sensitive to the detailed acoustic structure of the speech input. On the other hand, it is viewed as a non-linear dynamical system characterized by a limited number of stable states, or attractors, which allow the system to perform a discretization of perceptual space and which are associated with abstract perceptual categories. The categorization patterns observed partly derive from the mapping of the speech sounds onto a discrete and finite set of perceptual categories by means of a *continuous* potential function, as opposed to the sharp division between sounds and categories often posited. Unlike other hybrid approaches, however, the dynamical approach takes relative stability as a fundamental aspect of the categorization system. Gradient acoustic properties, frequency of occurrence of perceived categories, trajectory of speech sounds in the acoustic space, training, native language, and so on, affect the relative stability of categories and in turn, switching behavior. From this stems the wide range of phenomena observed, such as hysteresis, contrast, and bifurcations, that cannot be explained by an incremental averaging of exposures (statistical models that are based on interactive activation exhibit many of the same phenomena; McClelland & Vallabha, 2009).

Nonlinear dynamics offers an explanation of the bistability of speech perception, attributed to the coexistence of nearby but mutually exclusive attractor states in the perceptual space. In addition, theoretical and methodological tools (e.g., Erlhagen et al., 2006) are available that may allow the non-linear dynamical framework to be extended to the study of conversation interaction between speakers and to model the changing organization of perceived categories as this interaction unfolds in time.

Acknowledgments

This work was partly supported by the ACI Systèmes complexes en SHS Research Program (CNRS & French Ministry of Research) and by grants from the National Science Foundation (BCS-0414657 and BCS-0719683).

References

Case, P., Tuller, B., Ding, M., Kelso, J.A.S.: Evaluation of a dynamical model of speech perception. Perception & Psychophysics 57, 977–988 (1995)

Chistovich, L., Fant, G., de Serpa-Leitao, A., Tjernlund, P.: Mimicking of synthetic vowels. Quarterly Progress and Status Report, Speech Transmission Lab, Royal Institute of Technology, Stockholm, pp. 1–18 (1966)

Erlhagen, W., Mukovskiy, A., Bicho, E.: A dynamic model for action understanding and goal-directed imitation. Brain Research 1083, 174–188 (2006)

Gescheider, G.: Chapter 1: Psychophysical Measurement of Thresholds: Differential Sensitivity. In: Psychophysics: the fundamentals, 3rd edn. Lawrence Erlbaum Associates, Mahwah (1997)

Goldwater, S., Griffiths, T.L., Johnson, M.: A Bayesian framework for word segmentation: Exploring the effects of context. Cognition 112, 21–54 (2009)

Grossberg, S.: Resonant neural dynamics of speech perception. Journal of Phonetic 31, 423–445 (2003)

Hawkins, S.: Roles and representations of systematic fine phonetic detail in speech understanding. Journal of Phonetics 31, 373–405 (2003)

Hawkins, S.: Phonetic variation as communicative system: Perception of the particular and the abstract. In: Fougeron, C., D'Imperio, M., Kühnert, B., Vallée, N. (eds.) Papers in Laboratory Phonology X. Mouton de Gruyter, Berlin (2007)

Johnson, K.: Decisions and mechanisms in exemplar-based phonology. UC Berkeley Phonology Lab Annual Report, 289–311 (2005)

Kelso, J.A.S.: Dynamic Patterns: The Self-Organization of Brain and Behavior. MIT Press, Cambridge (1995)

Kent, R.D.: The imitation of synthesized vowels and some implications for speech memory. Phonetica 28, 1–25 (1973)

Kuhl, P.K.: A new view of language acquisition. Proceedings of the National Academy of Sciences 97, 11850–11857 (2000)

Ladefoged, P., Broadbent, D.E.: Information conveyed by vowels. Journal of the Acoustical Society of America 29, 98–104 (1957)

Lancia, L., Nguyen, N., Tuller, B.: Nonlinear dynamics of speech categorization: critical slowing down and critical fluctuations. The Journal of the Acoustical Society of America 123, 3077 (2008)

Liberman, A.M., Harris, K.S., Hoffman, H.S., Griffith, B.C.: The discrimination of speech sounds within and across phoneme boundaries. Journal of Experimental Psychology 54, 358–368 (1957)

Luce, P., McLennan, C.: Spoken word recognition: The challenge of variation. In: Pisoni, D., Remez, R. (eds.) The Handbook of Speech Perception, pp. 591–609. Blackwell, Malden (2005)

McClelland, J.L., Vallabha, G.: Connectionist models of development: Mechanistic dynamical models with emergent dynamical properties. In: Spencer, J.P., Thomas, M.S.C., McClelland, J.L. (eds.) Toward a unified theory of development: Connectionism and dynamic systems theory re-considered (2009)

McClelland, J.L., Mirman, D., Holt, L.L.: Are there interactive processes in speech perception? Trends in Cognitive Sciences 10, 363–369 (2006)

McLennan, C., Luce, P.: Examining the time course of indexical specificity effects in spoken word recognition. Journal of Experimental Psychology: Learning, Memory and Cognition 31, 306–321 (2005)

McMurray, B., Aslin, R.N., Toscano, J.C.: Statistical learning of phonetic categories: Insights from a computational approach. Developmental Science 12, 369–378 (2009)

Moore, B.C.J.: An introduction to the psychology of hearing, 5th edn. Emerald Group Publishing (2003)

Muchisky, M.M., Bingham, G.P.: Trajectory forms as a source of information about events. Perception & Psychophysics 64, 15–31 (2002)

Nguyen, N., Lancia, L., Bergounioux, M., Wauquier-Gravelines, S., Tuller, B.: Role of training and short-term context effects in the identification of /s/ and /st/ in French. In: Hazan, V., Iverson, P. (eds.) ISCA Workshop on Plasticity in Speech Perception (PSP2005), London, UK, pp. A38–A39 (2005)

Nguyen, N., Wauquier, S., Tuller, B.: The dynamical approach to speech perception: From fine phonetic detail to abstract phonological categories. In: Pellegrino, F., Marsico, E., Chitoran, I., Coupé, C. (eds.) Approaches to Phonological Complexity, pp. 193–217. Walter de Gruyter, Berlin (2009)

Öhman, S.E.G.: Coarticulation in VCV utterances: spectrographic measurements. Journal of the Acoustical Society of America 39, 151–168 (1966)

Pierrehumbert, J.: The next toolkit. Journal of Phonetics 34, 516–530 (2006)

Pisoni, D.B.: Auditory and phonetic memory codes in the discrimination of consonants and vowels. Perception & Psychophysics 13, 253–260 (1973)

Porter, R.J., Lubker, J.F.: Rapid reproduction of vowel-vowel sequences - Evidence for a fast and direct acoustic-motoric linkage in speech. Journal of Speech and Hearing Research 23, 593–602 (1980)

Remez, R.E., Rubin, P.E., Pisoni, D.B., Carrell, T.D.: Speech perception without traditional speech cues. Science 212, 947–950 (1981)

Repp, B.H., Liberman, A.M.: Phonetic category boundaries are flexible. In: Harnad, S.R. (ed.) Categorical Perception: The Groundwork of Cognition, pp. 89–112. Cambridge Univ. Press, Cambridge (1987)

Repp, B.H.: Categorical perception: Issues, methods and findings. In: Lass, N. (ed.) Speech and language. Advances in basic research and practice, vol. 10, pp. 244–335. Academic Press, Orlando (1984)

Repp, B.H., Crowder, R.G.: Stimulus order effects in vowel discrimination. Journal of the Acoustical Society of America 88, 2080–2090 (1990)

Repp, B.H., Williams, D.R.: Categorical trends in vowel imitation: Preliminary observations from a replication experiment. Speech Communication 4, 105–120 (1985)

Repp, B.H., Williams, D.R.: Categorical tendencies in imitating self-produced isolated vowels. Speech Communication 6, 1–14 (1987)

Repp, B.H., Healy, A.F., Crowder, R.G.: Categories and context in the perception of isolated steady-state vowels. Journal of Experimental Psychology: Human Perception and Performance 5, 129–145 (1979)

Shockley, K., Sabadini, L., Fowler, C.A.: Imitation in shadowing words. Perception & Psychophysics 66, 422–429 (2004)

Stevens, K.N.: Acoustic Phonetics. MIT Press, Cambridge (1998)

Stevens, S.S.: On the theory of scales of measurement. Science 103, 677–680 (1946)

Studdert-Kennedy, M.: Imitation and the emergence of segments. Phonetica 57, 275–283 (2000)

Tuller, B., Case, P., Ding, M., Kelso, J.A.S.: The nonlinear dynamics of speech categorization. Journal of Experimental Psychology: Human Perception and Performance 20, 1–14 (1994)

Tuller, B., Jantzen, M.G., Jirsa, V.K.: A dynamical approach to speech categorization: Two routes to learning. Invited contribution, New Ideas in Psychology 26, 208–226 (2008)

Vallabha, G.K., McClelland, J.L.: Success and failure of new speech category learning in adulthood: Consequences of learned Hebbian attractors in topographic maps. Cognitive, Affective and Behavioral Neuroscience 7, 53–73 (2007)

Vallabha, G.K., Tuller, B.: Perceptuomotor biases in vowel imitation. Journal of the Acoustical Society of America 116, 1184–1197 (2004)

Warren, R.M.: Perceptual restoration of missing speech sounds. Science 167, 392–393 (1970)

Warren, R.M.: Auditory illusions and the perceptual processing of speech. In: Lass, N.J. (ed.) Principles of Experimental Phonetics, pp. 435–466. Mosby, St. Louis (1996)

Warren, R.M., Warren, R.P.: Auditory illusions and confusions. Scientific America 223, 30–36 (1970)

A Neural Basis for Perceptual Dynamics

Howard S. Hock and Gregor Schöner

Abstract. Perceptual stability is ubiquitous in our everyday lives. Objects in the world may look somewhat different as the perceiver's viewpoint changes, but it is rare that their essential stability is lost and qualitatively different objects are perceived. In this chapter we examine the source of this stability based on the principle that perceptual experience is embodied in the neural activation of ensembles of detectors that respond selectively to the attributes of visual objects. Perceptual stability thereby depends on processes that stabilize neural activation. These include biophysical processes that stabilize the activation of individual neurons, and processes entailing excitatory and inhibitory interactions among ensembles of stimulated detectors that create the "detection instabilities" that ensure perceptual stability for near threshold stimulus attributes. It is shown for stimuli with two possible perceptual states that these stabilization processes are sufficient to account for spontaneous switching between percepts that differ in relative stability, and for the hysteresis observed when attribute values are continually increased or decreased.

The responsiveness of the visual system to changes in stimulation has been the focus of psychophysical, neurophysiological, and theoretical analyses of perception. Much less attention has been given to the role of persistence, the effect of the visual system's response to previous visual events (its prior state) on its response to the current visual input. Perceiving an object can facilitate its continued perception when a passing shadow briefly degrades its visibility, when attention is momentarily distracted by another object, when the eyes blink, or when a random fluctuation within the visual system potentially favors an alternative percept. Having perceived an object's shape from one viewpoint can facilitate its continued perception despite changes in viewpoint that distort its retinal projection, potentially creating a non-veridical percept. These examples highlight the importance of the visual system's prior state, not just for perceptual stability, but also for perceptual selection; i.e., for the determination of which among two or more alternatives is realized in perceptual experience.

In this essay we discuss three neural properties that form a sufficient basis for a theory of perceptual dynamics that addresses the relationship between persistence, responsiveness to changes in stimulation, and selection. These neural properties are: 1) Individual neurons have the intrinsic ability to stabilize their activation state. 2) Neurons responsive to sensory information (i.e., detectors) are organized

---

Howard S. Hock
The Department of Psychology and The Center for Complex Systems and Brain Sciences

into ensembles whose members respond preferentially to different values of the same attribute (e.g., motion direction). Members of such ensembles have overlapping tuning functions; i.e., a detector responding optimally to one stimulus value will also respond, though less strongly, to similar attribute values. 3) The activation levels of a detector affects and is affected by nonlinear excitatory and inhibitory interactions with other detectors.

On this basis, we examine the persistence of steady-state detector activation despite the presence of random perturbations, the effect of neural stabilization on a detector's response to stimulation, the crucial role of "detection instabilities" in minimizing perceptual instability and uncertainty for near-threshold stimuli, and the importance of differences in the rate-of-change in activation for perceptual selection. Finally, we demonstrate that the signature features of perceptual dynamics, spontaneous switching between percepts differing in relative stability, and hysteresis, follow from the same three neural properties.

1 Perceptual Stability: Natural or Otherwise

Natural, everyday percepts are almost invariably monostable. The same percept occurs each time a stimulus is presented. It rarely happens that two qualitatively different percepts are formed for the same stimulus (this would constitute bistability), and the experience of spontaneous switching between alternative percepts is likewise rare. Because everyday experiences of monostability are so pervasive, stability is not always recognized as an important perceptual property. Not so for James Gibson (1966), who attributed the stability of real-world percepts to the tuning of our visual system to unambiguous, invariant properties of stimulation.

Although Yuille and Kersten (2005) take a different position, maintaining that natural images are inherently ambiguous, they join Gibson (1966) and others in disdaining the usefulness of artificial stimuli for an understanding of perception in the natural environment. It is arguable, however, that many natural objects are potentially bistable (e.g., bumps and holes), but there is sufficient disambiguating contextual information in the natural environment to over-ride the potential of such objects to exhibit the dynamical behavior that is readily observed in the laboratory. Indeed, it is the exceptional situations accessible in the laboratory that most clearly bring the fundamental indeterminance of perceptual bistability into the domain of phenomenal perception.

Our dynamical research has taken place in well-controlled laboratory settings, where we have studied single-element apparent motion (Hock, Kogan & Espinoza, 1997; Hock, Gilroy & Harnett, 2002), displaced targets embedded in noise (Eastman & Hock, 1999), displaced rows of evenly spaced dots (Hock & Balz, 1994; Hock, Balz & Smollon, 1998; Hock, Park & Schöner, 2002), and the motion quartet (which is described below).

Single element displacements result in unique motion percepts, and many stimuli with multiple element displacements result in the perception of unique motion patterns. For instance, parallel horizontal motions are perceived for two vertically aligned elements alternating with two vertically aligned elements that are horizontally displaced, as in Figure 1a. This percept uniquely solves the motion correspondence problem; i.e., how visual elements presented during successive points in time are

A Neural Basis for Perceptual Dynamics

"paired" with respect to the start and end of perceived motion paths (Ullman, 1979). This is the case even though diagonal motions are in principle also possible for this stimulus. That is, despite single element motion being easily perceived for each independently presented diagonal displacement (Figures 1b and 1c), intersecting diagonal motions are never seen when the two diagonal displacements are combined in the same stimulus, as in Figure 1a.

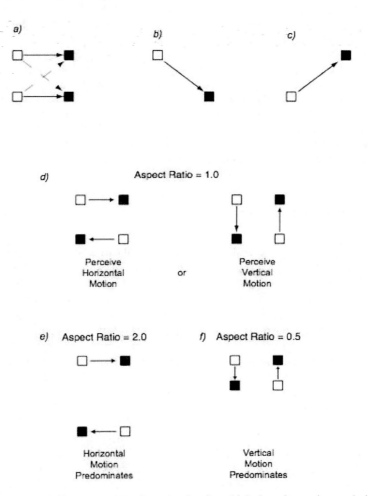

Fig. 1 (a) An illustrative apparent motion stimulus for which there is a unique solution to the motion correspondence problem (horizontal motion always is perceived), even though diagonal motions are possible (b and c). (d) The motion quartet, an apparent motion stimulus for which there are two qualitatively different solutions to the motion correspondence problem. For intermediate aspect ratios (the vertical divided by the horizontal distance between the elements), either horizontal or vertical motion is perceived. Horizontal motion predominates for relatively large aspect ratios (e) and vertical motion predominates for relatively small aspect ratios (f).

In contrast to the stimulus in Figure 1a, the motion quartet is an apparent motion stimulus for which two qualitatively different solutions can be realized in experience; either horizontal or vertical motion is perceived, the proportion of each depending on the aspect ratio of the quartet (Figures 1d-1f). Our research with this bistable stimulus has included psychophysical experiments and the dynamical modeling of spontaneous switching, hysteresis, selective adaptation, and activation-dependent detector interactions (Hock, Kelso & Schöner, 1993; Hock, Schöner & Hochstein, 1996; Hock, Schöner & Voss, 1997; Hock, Schöner & Giese, 2003; Hock, Bukowski, Nichols, Huisman, & Rivera, 2005; Hock & Ploeger, 2006; Nichols, Hock & Schöner, 2006). We currently are studying the effect of neural feedback on the stabilization of global motion patterns for stimuli composed of multiple motion quartets (Hock, Brownlow & Taler, in preparation).

It is for stimuli like the motion quartet that it is possible to directly observe the nonlinear mechanisms that bind stimulus specification with the ongoing neural activity resulting from preceding visual events, revealing fundamental properties of the processing mechanisms that are the basis for perception, not just in the laboratory, but in the natural environment as well. The most fundamental of these properties is neural self-stabilization.

2 Neural Stabilization

Whether an individual neural detector is activated by a stimulus or not, random events (perturbations) will cause its activation to fluctuate randomly with respect to some steady-state value. However, the variability of these fluctuations does not increase indefinitely over time. Although at first glance this may not be surprising, the "boundedness" of variability reflects a crucial, though often unrecognized feature of neural behavior, namely, that a neuron's activation is actively stabilized.

This idea can be made concrete by starting with any activation level for a neuron at any moment in time, and assuming that there is no interaction with other neurons. A random perturbation, if unconstrained, with equal probability will increase or decrease the neuron's activation. Assume it increases activation. The next and all following random perturbations will again with equal probability increase or decrease activation. Thus, there is nothing that systematically returns the activation from its increased level. Similarly, if an initial perturbation decreases activation, there is nothing that returns the activation from its decreased level. The same logic applies to any activation state generated by perturbations. Over time, states further and further removed from the initial activation state can be reached (e.g., by the chance event of a number of consecutive random increases in activation) and nothing drives the system systematically back from such states. It is intuitively clear, therefore, that the variance of activation would increase indefinitely over time. A formal argument of this kind led to an account for Brownian motion and the increase in time of the uncertainty about the location of a Brownian particle (Einstein, 1905).

The essential feature that keeps the variance of random fluctuations bounded is that successive random perturbations do not increase or decrease the neuron's activation level with equal probability. That is, the effects of random perturbations on activation are not unconstrained. When a random perturbation causes a fluctuation in activation, the change is opposed by the neuron's intrinsic ability to stabilize its activation, which reduces the size of the fluctuation. It is because of this resistance to the effects of random perturbations that there is an upper bound to the variance of random fluctuations in activation. The steady-state activation value of a neuron (or population of neurons) that is thus stabilized against the effects of random perturbations is referred to as an attractor.

2.1 The Biophysical Basis of Neural Stabilization

The biophysics of individual neurons provides a mechanism for achieving this stabilization of neural activation (Trappenberg, 2002). Specifically, the electrical potential across the membrane that separates the interior of a nerve cell from its inter-cellular environment is kept stable through the mechanisms of osmotic pressure. Ion pumps keep the concentration of different kinds of ions unequal on both sides of the membrane, the resulting flow of ions being in equilibrium when the electrical potential across the membrane just counterbalances the difference in ion concentration. If the equilibrium is perturbed (e.g., by an electrical current injected into the cell), the flow of ions quickly re-establishes the steady-state membrane potential.

With a neuron's membrane potential thus stabilized, synaptic input to the neuron increases the potential, increasing the probability that the neuron's activation will be transmitted to other neurons through action potentials traveling down its axon. In our account of neural dynamics (and most other such accounts) the stabilized membrane potential, averaged over local neural populations composed of hundreds or thousands of individual neurons, is sufficient to account for the mapping of psychophysical events onto patterns of neural activation. To be sure, the mathematical relationship between ion flows that stabilize the membrane potential of individual neurons and the stability properties of neural populations is not well understood. Eggert and van Hemmen (2001) have provided one such account, but it is limited by the simplifying assumptions that the constituents of a neural population are both identical in their responsiveness to stimulation and non-interactive. This notwithstanding, it is reasonable to proceed based on the principle that stability properties of neural populations are inherited from the dynamics through which individual neurons stabilize their membrane potential (Jancke, Erlhagen, Dinse, Akhavan, Giese, Steinhage & Schöner, 1999).

2.2 The Time Scale

The extent to which a neuron or population of neurons resists fluctuations in activation caused by random perturbations depends on how quickly fluctuation-opposing changes emerge within the neurons. This determines the time scale of stabilization. If there were only one instantaneous perturbation, the pre-perturbation activation level (i.e., the average membrane potential) would be restored over an interval determined by the time scale. This is called the "relaxation time." However, random perturbations occur continually, so depending on the time scale, there is sufficient time only for the partial restoration of the fluctuation in activation caused by one perturbation before the next one occurs. The faster the time scale, T, the greater the restoration of activation, and therefore, the greater the resistance to the effects of the random perturbation.

2.3 The Core Dynamical Concept

Neural stabilization provides the basis for the core concept of a theory of perceptual dynamics. That is, whatever causes a change in the current neural activation, u, will be opposed in the immediate future by a change in activation, du/dt, in the opposite direction. Activation increases in the immediate future when the change in activation, du/dt, is positive (because u has decreased) and it decreases in the immediate future when the change in activation, du/dt, is negative (because u has increased). This relationship among current levels of activation, u, and changes in activation that will occur in the immediate future (du/dt) can be expressed as:

$$du/dt = -u/T$$

where T is the time scale of perturbation-opposing reactions within the neuron. T determines the size of the change in activation in opposition to random fluctuations, with larger compensating changes (larger values of du/dt) occurring when T is smaller/faster.

2.4 Stable Activation States in the Absence of Stimulation

In classical, non-dynamical approaches to the study of perception, unstimulated detectors are simply inactive, and although there are numerous dynamical accounts of perception, they generally do not address the status of detectors when they are unstimulated. In our dynamical conceptualization, however, a neuron's ability to stabilize its activation means that populations of detectors have stable activation states even when they are unstimulated, and irrespective of their connectivity to other detectors. This means that unstimulated detectors can maintain activation near an attractor value that is below the threshold level required for perception, thereby minimizing the likelihood that random fluctuations would cause the activation of unstimulated detectors to rise above this threshold. The stabilization of activation in the absence of stimulation (i.e., in the vicinity of the no-stimulus, resting level) can be characterized by adding h to the dynamical equation:

$$du/dt = (-u + h)/T$$

where h is the detector's resting level.

A Neural Basis for Perceptual Dynamics

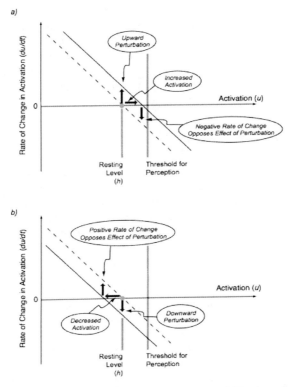

Fig. 2 The straight lines with negative slope represent the stabilization of activation (u) in the absence of stimulation (i.e., in the vicinity of the no-stimulus, resting level). This is determined by the dynamical equation: $du/dt = (-u + h) / T$, where the rate-of-change, du/dt, determines whether and by how much activation will increase or decrease in the immediate future. h is the detector's resting level and T is the time scale of activation change. Because of the negative slope, random fluctuations in activation are resisted by changes in the opposite direction, with activation stabilizing at the attractor for the no-stimulus/resting state ($u^* = h$). Panels (a) and (b) differ with respect to whether the perturbation increases or decrease activation

As can be seen in the graphical representation of the equation (Figure 2), the attractor for the no-stimulus (resting) state is the activation value $u^* = h$. By simple calculation from the above equation, it is the steady-state activation level when the rate-of-change, du/dt, is 0. When a perturbation occurs, it imposes a randomly determined upward or downward rate of change in activation; i.e., it creates a tendency for activation to change in a particular direction. Such perturbations cannot be instantaneous. They must last long enough for activation to reach a value different from h. It can be seen in Figure 2a that the value of du/dt is negative when the perturbation imposes a positive rate of change, so neural stabilization opposes the effect of the perturbation by proportionally decreasing activation following the perturbation. It similarly can be seen in Figure 2b that the value of du/dt is positive when a random perturbation imposes a negative rate of change in activation, so neural stabilization opposes the effect of the perturbation by proportionally increasing activation after the perturbation. Irrespective of direction, the larger the change in activation caused by the perturbation, the greater the opposing change, du/dt. The latter, together with randomness with respect to whether perturbations have positive or negative effects, stabilizes activation in the vicinity of the resting level.

2.5 Response to Stimulation

What happens when a stimulus is presented for which detectors are responsive? In classical non-dynamical approaches, the steady-state activation level of a detector is determined by the strength of the stimulus, and there is little concern with how activation evolves over time toward these steady-state values. In our dynamical conceptualization, however, the initial response to the presentation of a detector-activating stimulus occurs in the context of activation states (at the resting level) that are stabilized with respect to the effects of random perturbations. This stabilization mechanism therefore determines how activation evolves from the resting level toward the steady-state activation level determined by the stimulus.

This can be made intuitive by imagining that a stimulus presented for a finite period of time is equivalent to a dense sequence of activation-increasing perturbations (rather than a random sequence of positive and negative perturbations, as in the preceding section). Each of these "perturbation-induced" upward fluctuations in activation is partially opposed by neural stabilization, so at a rate determined by the time scale, T, of the neural stabilization mechanism, successive excitatory perturbations incrementally move a detector's activation away from its (no-stimulus) resting level, toward the stimulus-determined activation level. By imagining the stimulus as a dense sequence of partially restored excitatory fluctuations, it can be understood that the time course of the activation as it rises from the resting level depends on the same neural stabilization mechanism that keeps the activation of unstimulated detectors from randomly fluctuating above the threshold level for perception. The activational effect of a stimulus on a detector ensemble therefore can be characterized by adding S to the previously introduced dynamical equation:

$$du/dt = (-u + h + S) / T$$

When the stimulus is presented, it imposes a positive rate-of-change on activation; i.e., activation increases immediately after the stimulus is presented, which moves it from the no-stimulus attractor, $u^* = h$, toward the stimulus-determined attractor, $u^* = h + S$. (It can be seen from the graphs in Figure 3 that du/dt is positive in relation to the stimulus-determined attractor when activation has been at the resting level.) A comparison of Figures 3a and 3b shows that the time scale determines how quickly activation changes as it moves away from the resting level, toward the stimulus-determined attractor; smaller/faster time scales result in larger, more rapid shifts in activation toward the attractor.

A Neural Basis for Perceptual Dynamics

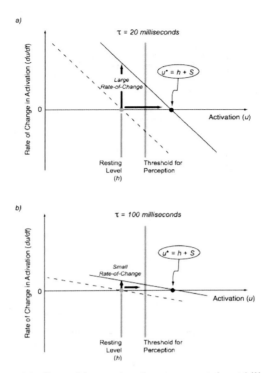

Fig. 3 The broken straight lines with negative slope represent the stabilization of activation (u) in the absence of stimulation, as in Figure 2, and the parallel solid lines represent the stabilization of activation (u) in the presence of stimulation. The latter is determined by the dynamical equation: $du/dt = (-u + h + S) / T$, where the rate-of-change, du/dt, determines whether and by how much activation will increase of decrease in the immediate future. h is the detector's resting level, T is the time scale of activation change, and S is the stimulus-initiated activation. Activation increases to the steady-state attractor value, $u^* = h + S$, at a rate determined by the time scale of the dynamics, which differs in panels (a) and (b).

This evolution of activation for the stimulated detector is illustrated in Figure 4 for two different time scales, which shows that there is a greater rate-of-change in activation for the faster time scale as activation rises from the resting level. In addition, random fluctuations are less variable (with the same level of random noise perturbations) for the faster time scale. So long as the detector's activation is not influenced by interaction with other detectors (or adaptation), activation would settle near the attractor, $u^* = h + S$, for both time scales. For the simulations in Figure 4a, $h = -8$ and $S = 16$, so it is readily calculated from the above equation that when du/dt is zero, the attractor is at $u^* = -8+16 = 8$.

It is important to note that the comparisons in Figures 3 and 4 are made in order to provide an intuitive understanding of the effects of time scale on changes in activation. At this juncture, it is not possible to directly determine time scales for perceptual dynamics (slower time scales cannot be discriminated from higher levels of noise) or to separate contributions to time scale from the stabilization of

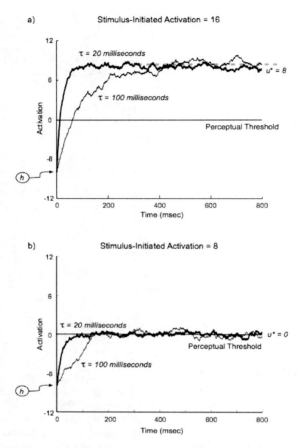

Fig. 4 This evolution of activation as the result of stimulation for two time scales, demonstrating the faster change in activation for the briefer time scale. Activation settles near the attractor, $u^* = h + S$, for both time scales, with $S = 16$ in panel (a) and $S = 8$ in panel (b). Activation for the latter would straddle the threshold for perception, $u^* = -8 + 8 = 0$, so even small random fluctuations would render perception highly unstable.

membrane potentials for individual neurons, and from to-be-discussed interactions between detectors that compose neural ensembles.

2.6 Perceptual Thresholds, Perceptual Stability and Uncertainty

In accounts of perception based on neural dynamics, it is assumed that a stimulus attribute is perceived when the average activation of detector populations responsive to that attribute is stabilized at a level that exceeds a particular threshold value (perhaps determined by the level of membrane potential that results in the transmission of action potentials). However, neural stability does not

A Neural Basis for Perceptual Dynamics

guarantee perceptual stability. If the stimulus-initiated activation in the preceding example were weaker ($S = 8$), the stabilized activation (i.e., the attractor) would straddle the threshold for perception, $u^* = -8+8 = 0$. As a result, even small random perturbations would rapidly shift activation back and forth across the threshold, rendering perception highly unstable, and therefore, highly uncertain (Figure 4b).

This kind of near-threshold uncertainty, which is classically the domain of signal detection theory (Green & Swets, 1966), parallels the dynamical account presented thus far. That is, both entail the detection of a signal (the steady-state activation level) embedded in noise (random fluctuations in activation), and a criterion (the perceptual threshold) that determine whether or not the attribute is present (perceived). While signal detection theory is sufficient to account for the near-threshold uncertainty that occurs in many contexts, near-threshold uncertainty does not generally obtain in motion perception. For example, Hock, Kogan and Espinoza (1997) found values of luminance contrast change that result in the perception of single-element apparent motion for half the trials and the perception of nonmotion for the other half. For almost all the trials, either motion or nonmotion was clearly perceived. There were only occasional trials for which subjects were uncertain regarding what they perceived. Consistent with such experimental results, perceptual uncertainty does not occur in our dynamical conceptualization because as activation rises from the resting level for individual detectors, interaction with other detectors emerges, and what would be near-threshold activation levels for the individual detectors are boosted to above-threshold levels and further stabilized at those levels. How this occurs is discussed next.

3 The Stabilization of Activation within Detector Ensembles

Up until this point in the discussion we have focused on the stabilization properties of individual detectors (or populations of detectors) in the absence of interaction among them. In this section we step back to establish the neuro-anatomical basis for detector interactions and the particular interactions that are the basis for detection instabilities.

3.1 Neural Connectivity

In a neural network conceptualization of perception, stimulus specification is roughly characterized by the feedforward path through the network. Activation induced in this path is largely stimulus determined, the preferential responding of different detectors occurring by virtue of their receptive fields being structured to realize various feature extraction filters (for motion, orientation, line length, texture, color, etc.). However, most neuronal activity entails more than the feedforward stream. Braitenberg (1978) has estimated that 95% of the input to each cortical neuron comes from its connectivity with other cortical neurons, and Felleman and Van Essen (1991) have determined that there is more feedback than feedforward connections between higher- and lower-level areas in the brain.

Given this neuro-anatomical evidence, one cannot expect the visual system to simply "compute" the perceptual output from the stimulus input. When a stimulus is presented, it is necessary to also take into account what already is occurring within highly interconnected neural networks. This is the domain of neural dynamics, which was first introduced into the study of perception by Stephen Grossberg and his colleagues (Grossberg, 1973; Grossberg & Mingolla, 1985; Grossberg & Rudd, 1992; Francis, & Grossberg, 1996; Chey, Grossberg, & Mingolla, 1997; Baloch & Grossberg, 1997). In our analysis, perceptual dynamics describe how activation within a network of detectors evolves in time under the influence of both current input (stimuli consistent with the preferential responding of detectors in the network) and ongoing activity (persistent activation due to earlier stimulus input).

3.2 Interaction

The most important consequence of neural connectivity is that a detector, when activated, can influence the activation levels of other detectors, either by increasing their activation through excitatory interaction or decreasing their activation through inhibitory interaction. An essential feature of such interactions is that they are activation dependent. That is, the more strongly a detector is activated, the greater its interactive influence on the detectors with which it is connected. This is illustrated by the sigmoidal function in Figure 8b.

Thus, each detector in an interconnected network is subject not only to the time-varying activational effect of the stimulus and random perturbations that produce fluctuations in its activation, but also to activation-dependent interactive influences from other detectors. (Another kind of influence, which we will ignore in this chapter, is activation-dependent adaptation; but see Hock et al. 2003; Nichols et al. 2006.) Because of this interaction, all contributions to a detector's activation change its influence on other detectors, which ultimately comes around again by affecting their interactive influence on the detector whose activation is the source of the interaction. This ongoing re-cycling of change, or recurrence, reflects the state-dependence of the network; i.e., the evolving activation state of a detector depends on its own previous activation state as well as the activation state of the detectors with which it interacts. In this way, the presentation of a stimulus initiates a recurrent cycle of activation change in networks of detectors that moves activation toward steady-state values, values for which rates-of-change in activation are near zero.

3.3 Interaction within Detector Ensembles

As indicated earlier, a detector's activation can be stabilized at a steady-state value, but this will not result in perceptual stability if the steady-state value is near the threshold for perception. Even small random perturbations would then be sufficient to rapidly shift activation back and forth across the threshold, rendering perception unstable and uncertain. This "problem" can be resolved by assuming that perception requires the stimulus-initiated activation of individual detectors, but that it is ultimately determined by the pattern of activation within ensembles of interacting detectors that respond preferentially to different values of the same attribute. Because these detectors have overlapping tuning functions, some will respond optimally to one attribute value, others will respond, though less strongly, to that attribute value, and still others will not respond at all to that attribute value. It is within the population of stimulated detectors that excitatory and inhibitory interactions create detection instabilities.

3.4 Detection Instability

When an appropriate stimulus is presented, the activation of responding ensemble members rises from their resting levels (at a rate determined by each detector's neural stabilization mechanism), and an activation level is approached for which the population of detectors will boost each others' activation through mutual excitatory interaction. Up until this point, activation is still subthreshold for perception, so whether or not further growth in stimulus-initiated activation engages excitatory interactions will have dramatic effects on activation levels for the population of stimulated ensemble members. If stimulus-initiated activation is insufficient to initiate excitatory interactions among ensemble members, activation for the stimulated detectors will be stabilized below the threshold level required for perception. However, if stimulus-initiated activation is sufficient to initiate excitatory interactions, detectors that respond optimally will increase the activation of detectors that prefer similar, but different attribute values. And the latter, in turn, will increase the activation of the detector already responding optimally to the stimulus.

This reciprocal excitation will result in detector activation rapidly passing through a "detection instability" (Bicho, Mallet, & Schöner, 2000; Schöner, 2008; Schneegans & Schöner, 2008). Activation will accelerate well past levels that are near-threshold for perception because detector interactions are activation dependent; i.e., the more strongly a detector is activated the greater its excitatory effect on other ensemble members. The indefinite increase in activation for these self-excited detectors is prevented by the accompanying rise in activation-dependent inhibitory interactions, resulting in activation settling at a stable value (attractor) for the population of detectors that respond preferentially to attribute values at or near to the attribute value of the stimulus. The inhibitory interactions that limit the growth in self-excited activation, though weaker than the excitatory interactions, come from ensemble members with preferences for a wider range of attribute values (Amari, 1977; Wilson & Cowan, 1973). As a result, pairs of detectors with similar preferences will more strongly excite than inhibit each other, whereas the reverse will be the case for pairs of detectors with dissimilar preferences.

Inhibitory interactions are thus comparatively long-range. They can be effective over longer distances (in attribute space) than short-range excitatory interactions, so they prevent the spread of activation to inter-connected detectors with preferences for attribute values that are much different than the attribute value of the stimulus. It will be seen in Section 4 that this long-range inhibition also is important for perceptual selection, which comes into the picture when stimulus-initiated activation for much different attribute values makes it possible for the activation of more than one population of ensemble members to pass through a detection instability.

3.5 Stabilization of Activation within Detector Ensembles

When a self-excited population of detectors passes through a detection instability, its activation is boosted well past the threshold level required for perception, and in addition, its stability is enhanced with respect to the fluctuations in activation produced by random perturbations. The reason is as follows. If the net effect of random perturbations over the population of detectors is for activation to fluctuate upwards, the increased activation will spread through detector pairs with similar preferences via short-range excitatory interactions. However, detector activation due to the net-upward fluctuations is compensated for by increased inhibitory interactions. In the case of inhibition there is both short-range inhibition between pairs of detectors with similar preferences, and long-range inhibition between pairs of detectors with dissimilar preferences. The net effect will be to oppose the net-increase in activation by increasing activation-reducing inhibitory interactions.

The opposite occurs when random perturbations result in a net downward fluctuation in activation for the population of self-excited detectors. The reduced activation will spread through the reduction in short-range excitation between detector pairs with similar preferences. However, the reduction in activation also will decrease both short-range inhibition between pairs of detectors with similar preferences, and long-range inhibition between pairs of detectors with dissimilar preferences. The net effect will be for the population of self-excited detectors to oppose the net-decrease in activation by decreasing activation-reducing inhibition, which effectively increases activation.

Stability therefore is asymmetrical for the two states associated with the detection instability. If stimulus-initiated activation is too weak to engage the excitatory interactions that boost the ensemble's activation, activation will remain below the threshold level required for perception. Stability in the presence of random perturbations then would be determined by the intrinsic neural stabilization properties of the stimulated detectors (i.e., by the balance of membrane potentials and ion concentrations). If stimulus-initiated activation is sufficient for activation to produce a self-excitatory boost in activation, stability in the presence of random perturbations would then be determined by both the neural stabilization properties of the stimulated detectors and the balance of excitatory and inhibitory interactions that results from their passing through a detection instability, as described above.

3.6 Removing the Stimulus

The new stable state induced by the detection instability resists decay when the activation-initiating stimulus is removed. This is because excitatory interactions keep boosting detector activation even as the removal of the stimulus induces a reduction in activation. Although this resistance to the removal of the stimulus ultimately fails, it slows the return of activation to the detectors' no-stimulus resting levels, so that activation persists for time intervals that are much longer than those needed to initially stimulate the detectors and boost activation through the detection instability. This lingering of activation (below the threshold for perception, but above the resting level) potentially accounts for percepts exhibiting stability over surprisingly long temporal intervals (Leopold, Wilke, Maier & Logothetis, 2002).

4 Perceptual Selection

Thus far we have discussed how activation within a detector ensemble is stabilized around detectors that optimally respond to a particular attribute value. The principle is that above-threshold activation for these self-excited detectors would signify the perception of this attribute value. However, in natural perception (and in the laboratory), activation can be simultaneously initiated around more than one attribute value. For example, both horizontal and vertical motions are simultaneously stimulated by the motion quartet described in Figures 1d-1f. When the detectors responsive to one attribute value are more strongly stimulated than the detectors responsive to other attribute values, the long-range inhibitory interactions that stabilize activation in one part of the attribute space (Sections 3.3 and 3.4) can reduce stimulus-initiated activation in other parts of the attribute space, preventing activation at that location from passing through an activation-boosting detection instability. This can result in one overwhelmingly dominant percept (the horizontal motion in Figure 1a), even though the alternative attribute values (the diagonal motions in Figures 1b and 1c) would be perceived in the absence of competition from the detectors that respond optimally to horizontal motion. The characteristics of this inhibitory competition, which would contribute to the monostability typical of natural perception, is best studied in the case of bistability; i.e., when the competing stimulus attributes are relatively similar with respect to the level of activation that they stimulate.

4.1 Perceptual Bistability

Perception is said to be bistable when two different percepts are possible for the same stimulus. When the competing percepts are based on different values of the same attribute (we will call the values A and B), the value selected for perception depends on activation-dependent inhibitory interactions between the populations of detectors responsible for the perception of each attribute value. When the stimulus-initiated activations for A and B are equal, random fluctuations occurring as activation rises

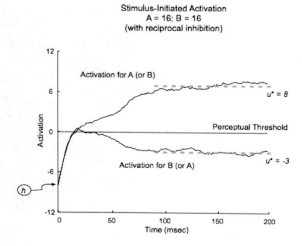

Fig. 5 When the stimulus-initiated activations for attributes A and B are equal, random fluctuations occurring as activation rises from the resting level result in detector populations that preferentially respond to one attribute having a momentary activational advantage over detector populations that preferentially respond to the other attribute. Reciprocal, activation-dependent inhibitory interactions between the two populations drive their activations apart such that the randomly advantaged attribute has its activation stabilized at an above-threshold level ($u^* = 8$) while its competitor is suppressed to a sub-threshold activation level ($u^* = -3$).

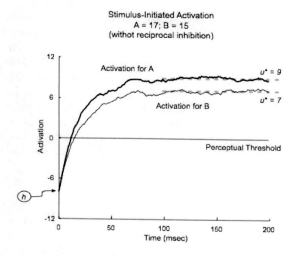

Fig. 6 When the stimulus-initiated activations for attributes A and B are unequal, and there are no activation-dependent inhibitory interactions between the detector populations responsive to these attributes, both would reach levels of activation that are above the threshold value required for perception, so both would be perceived simultaneously.

A Neural Basis for Perceptual Dynamics

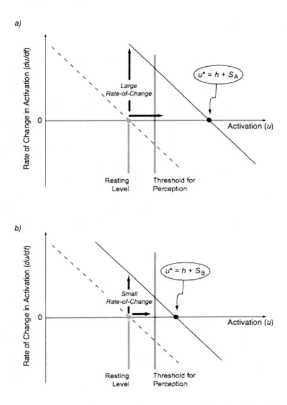

Fig. 7 Graphical representations for the stabilization of activation for non-interactive detector populations responsive to attributes A (panel a) and B (panel b), which differ with respect to their stimulus-initiated activation, as in Figure 6. Activation departs more quickly from the resting level (h) for the stronger (S_A) compared with the weaker attribute (S_B). This follows from the rate-of-change, du/dt, being larger for attribute A because its new attractor ($u^* = h + S_A$) is further from the initial, no-stimulus attractor ($u^* = h$) compared with the new attractor for B ($u^* = h + S_B$).

from the resting level result in detectors that preferentially respond to one attribute value having a momentary activational advantage over detectors that preferentially respond to the other attribute value, enabling the former to inhibit the latter more than vice versa (Figure 5). As activation increases, reciprocal, activation-dependent inhibition drives their activation levels further and further apart, with activation stabilizing at an above threshold level for one alternative and at a subthreshold level for the other. In this way, random fluctuations in activation lead to the perceptual selection of one of the equally stimulated alternatives. (In all the simulations, the inhibitory interaction has a maximum value of 11. With $S_A = S_B = 16$ for the simulation in Figure 5, the attractor for the above-threshold attractor is near $u^* = -8+16 = 8$, and the attractor for the subthreshold attractor is near $u^* = -8+16-11 = -3$.)

4.2 Rates of Change in Activation

We next examine selection between competing attribute values, A and B, when the detectors that respond preferentially to A are more strongly stimulated than the detectors that respond preferentially to B. If there were no activation-dependent inhibitory interactions between these detector populations, both would reach levels of activation that are above the threshold value required for perception. Both would be simultaneously perceived, which is contrary to true bistability. This is illustrated by trajectories (without interaction) for the evolution of activation toward the attractors for A, $u_A^* = h + S_A$, and for B, $u_B^* = h + S_B$, are indicated in Figure 6. (With $S_A = 17$ and $S_B = 15$, $u_A^* = -8+17 = 9$ and $u_B^* = -8+15 = 7$.) It can be seen in Figure 6 that upon stimulus presentation, the rate of change of activation as it rises from the no-stimulus resting state is greater for the detectors responding preferentially to A than for the detectors responding preferentially to B. This is shown in a different way by the graphical representations in Figure 7.

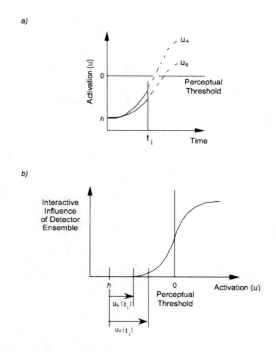

Fig. 8 When there are reciprocal, activation-dependent inhibitory interactions between detector populations, differences in the rate-of-change of activation (panel a) are of critical importance. As illustrated in panel b, this is because the detector population (for attribute A), with the faster rate-of-change, will reach an activation level where it will begin to inhibit its competing detector population (for attribute B), with a slower rate-of-change, before there can be an inhibitory influence in the opposite direction.

That is, activation for the stronger stimulus departs more quickly from the resting level, h, because the value of du/dt is greater when the new attractor is further from the resting level.

4.3 Perceptual Selection of the Favored Stimulus Alternative

As illustrated in Figure 8, the key to the perception of attribute value A when it is favored by the stimulus is that its detectors reach an activation level where they can begin inhibiting the activation of the detectors for attribute value B before the reverse occurs. This is why differences in the rate-of-change of activation are of critical importance. As a consequence of its faster rate-of-change, the activation for A increases to its above-threshold attractor value near $u^* = h + S_A = -8+17 = 9$), while at the same time activation for B is suppressed by inhibition from A, stabilizing below the perception threshold at attractor value near $u^* = h + S_B - I_A = -8+15-11 = -4$ (Figure 9). A comparison of Figures 6 and 9 shows that the reciprocal inhibitory interaction substantially separates the alternative activation states, facilitating the perceptual selection of the alternative that is most strongly specified by the stimulus. (Note: The simulations in this section and the remainder of the essay do not include the additional divergence in activation due to the self-excitation that occurs when the more strongly activated alternative passes through a detection instability.)

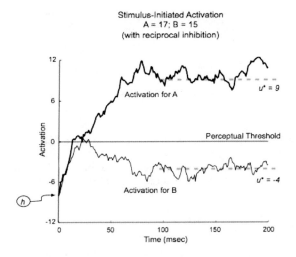

Fig. 9 The evolution of activation in the presence of reciprocal, activation-dependent inhibitory interactions when the more strongly activated detector population (for attribute A) suppresses the activation of its competitor, the detector population for attribute B.

4.4 Perceptual Selection of the Unfavored Stimulus Alternative

The detectors for A are more strongly stimulated than the detectors for B, so its perception is signified most of the time when the stimulus is presented. However, bistability means that B also can be perceived, though not necessarily as often as A. As illustrated by the activation trajectories in Figure 10, a sufficiently large fluctuation can result in B's activation becoming larger than A's as activation for both increase from the no-stimulus, resting level, h. B then can begin inhibiting its stimulus-favored competitor, A, before the reverse occurs. B's activation would then rise to an attractor value near $u_B^* = h + S_B = -8+15 = 7$, which lies above the threshold for perception, while A's activation decreases toward an attractor value near $u_A^* = h + S_A - I_B = -8+17-11 = -2$, which is subthreshold for perception. The perceptual selection of B is signified, but because this requires a random fluctuation that reverses the relative activation of the detectors for A and B as they rise from the resting level, it is perceived less often than A.

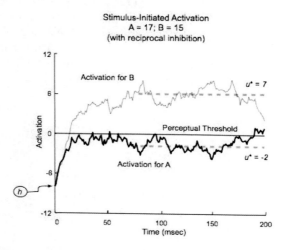

Fig. 10 The evolution of activation in the presence of reciprocal, activation-dependent inhibitory interactions when the more weakly activated detector population (for attribute B) gains a momentary activational advantage as a result of random fluctuations in activation, and suppresses the activation of its competitor, the detector population for attribute A.

5 Objects and Feature Integration

We have focused in this essay on the dynamical basis for the perception of individual attributes, arguing that it is determined by the pattern of activation over an ensemble of detectors responsive to different values of the same attribute. However, natural objects and even artificial objects created in the laboratory are multi-dimensional. The integration of the attributes (or features) belonging to the same object, called the binding problem, has been the subject of numerous empirical and computational studies (see Treisman, 1998). The dynamical framework described in this essay suggests that short-range excitatory interactions among attribute/feature ensembles at similar retinal locations might be sufficient to account for their integration. Alternatively, higher-level object units or units in working memory might receive input from attribute ensembles at similar retinal locations, and in turn provide feedback to those ensembles that maintains precise spatial information and affects activation within each attribute ensemble in a manner consistent with the activated higher-level units (Johnson, Spencer & Schöner, 2008).

Evidence consistent with the latter alternative comes from a recent study with stimuli composed of four motion quartets, as illustrated in Figure 11 (Hock, Brownlow & Taler, in preparation). For this arrangement, the motion directions for the individual quartets are integrated into a global rotational pattern (alternating clockwise and counterclockwise rocking motion; the rotation of the outer elements predominates), likely due to the activation of global motion detectors in Area MSTd (Tanaka & Saito, 1989). Feedback from the global to motion detector ensembles with different directional preferences was indicated by the motion for the quartets being in directions consistent with global rotation. For example, horizontal motion for the quartets on the top and bottom of the configuration even though their aspect ratio would otherwise favor motion in vertical directions, as in Figure 1f. This kind of result is consistent with Treisman's (1998) prediction that "The strongest evidence will come when changes in neural activity are found to coincide with perceived changes in binding, perhaps in ambiguous figures or attentional capture (page 35)."

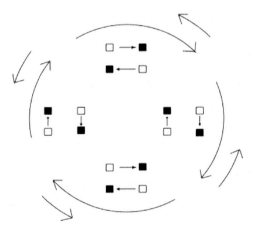

Fig. 11 Stimuli composed of four motion quartets arranged in a configuration for which the motion directions for the individual quartets can be integrated into a global rotational rocking pattern (alternating clockwise and counterclockwise rotation, with the perception of rotation predominating for the outer elements).

6 The Signature Features of Perceptual Dynamics

The stated purpose of this essay was to show that three neural properties are sufficient to provide the basis for a theory of perceptual dynamics that accounts for the relationship between persistence, responsiveness to stimulation, and selection. We have shown how perceptual persistence results from the intrinsic neural stabilization of individual detectors (through the balance of the membrane potential and ion concentrations; Section 2.1) and from the balance of excitatory/inhibitory interactions within a population of self-excited detectors with preferences for similar attribute values (Section 3.4). We have shown how change-resistant neural mechanisms affect the rate at which activation changes in response to changes in stimulation (Section 2.5), and how near-threshold perceptual uncertainty is minimized when stimulus-initiated activation is amplified by self-excitation (Section 3.3). Finally, we have shown how bistability arises as a result of reciprocal, activation-dependent inhibitory interactions between populations of detectors that preferentially respond to different values of the same attribute, and why differences in the rate-of-change activation are critical for the perceptual selection of the alternative percept that is most strongly specified by the stimulus (Sections 4.2 and 4.3). We show in this section that a perceptual dynamics based on these neural properties can result in the signature features of a dynamical system: spontaneous switching between percepts differing in relative stability, and hysteresis.

6.1 *Spontaneous Switching*

When two percepts are possible for the same stimulus, sufficiently large random perturbations can cause spontaneous switches between the percepts. The probability of a switch is inversely related to the likelihood of perceiving one of the alternatives when the bistable stimulus is presented. That is, the more likely it is for a percept to occur when a stimulus is presented, the less likely it is that there will be a spontaneous switch to the alternative percept.

This is illustrated by our earlier example in which stimulus-initiated activation is greater for the detectors responsible for the perception of A than for the detectors responsible for the perception of B. A is more likely to be perceived because it is more strongly specified by the stimulus, but sometimes B is perceived instead. Comparing Figures 9 and 10, it can be seen that the activational difference between the perceived and unperceived (subthreshold) alternatives is greater when the more likely of the alternatives is perceived (Figure 9) than when the less likely of the alternatives is perceived (Figure 10). A larger, less probable perturbation would be required to overcome the larger activational advantage of the perceived alternative when it is favored by the stimulus (A) compared with when it is not favored by the stimulus (B).

The relative stability of the two percepts is further illustrated in Figure 12 by continuing the simulation for a longer period of time, which provides more opportunity for the occurrence of random perturbations large enough to produce

A Neural Basis for Perceptual Dynamics

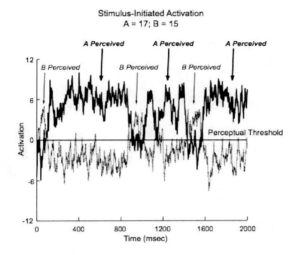

Fig. 12 A simulation demonstrating the relative stability of the percepts for which attributes A and B are perceived. Stimulus-initiated activation is greater for attribute A, so its perception is more stable than the perception of attribute B. This is indicated by the longer temporal intervals over which attribute A is perceived compared with the perception of attribute B.

perceptual switches (noise strength was increased in order to increase the frequency of switching). As is evident in Figure 12, both A and B are perceived, but the temporal intervals over which A is perceived are much longer than the temporal intervals over which B is perceived. The perception of A is more stable.

6.2 Hysteresis

Persistence (or its lack) in the presence of non-systematic, random events (passing shadows, distractors of attention, eye blinks, neural fluctuations) is observed dynamically in Section 6.1 as the dependence of spontaneous perceptual switching on the relative stability of the alternative percepts that are possible at a given moment. Such switching is rare in the natural environment because of the strong dominance of one percept, and because of the presence of disambiguating contextual information.

Persistence also facilitates the continuation of a previously established percept despite systematic changes in stimulus input, as might occur when the retinal projection of an object is distorted by gradual changes in viewpoint (due to the motion of an object or the egomotion of the perceiver). Objects in the world are invariant despite changes in viewpoint, so it is of obvious benefit to maintain the percept of an object that is established with less distorting projection angles.

To frame this in a manner consistent with our earlier examples, assume that perceiving the veridical shape of an object depends on the perception of attribute value A, but changes in viewpoint distort the retinal projection, favoring the

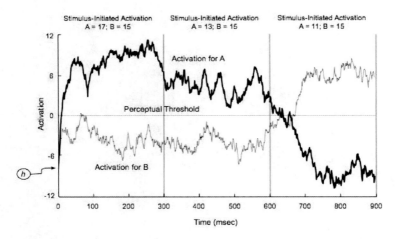

Fig. 13 A simulation demonstrating hysteresis. The perception of attribute A is established during the first 300 millisecond interval, when it is the stronger of the two attributes. Hysteresis is indicated by the perception of attribute A persisting during the second 300 millisecond interval despite the stimulus changing such that B becomes the stronger attribute. It is only during the final 300 millisecond interval, when the perception of attribute B is even more strongly favored by the stimulus, that there is a switch to its perception.

perception of attribute value B, and thus, a different, nonveridical shape. Hysteresis, the persistence of percept A despite systematic changes in the relative strength of attributes A and B, is illustrated in Figure 13. It can be seen in the figure that the perception of A, established when it is the stronger of the two attributes (during the first 300 millisecond interval of the simulation), is maintained despite the stimulus changing, and B becoming the stronger attribute (during the second 300 millisecond interval of the simulation). It is only when the advantage of stimulus-initiated activation more strongly favors attribute B that the initial perception of A gives way to the perception of B (during the final 300 millisecond interval of the simulation).

7 Conclusion

Over the years, there have been many physics- and mathematics-based entry points into the study of perceptual dynamics (e.g., Ditzinger & Haken, 1990; Poston & Stewart, 1978). As indicated earlier, Grossberg and his colleagues were the first to specifically study brain functions in terms of dynamics, building complex neural architectures in order to account for a wide range of perceptual and cognitive phenomena. (See also Wilson & Kim, 1994.) We have taken a different approach in this essay and in an earlier article (Hock, Schöner & Giese, 2002). That is, we have shown that a few basic neural properties are sufficient to provide the foundation for an understanding of the dynamical characteristics of perception, characteristics often over-looked or "taken for granted" by most investigators of perceptual phenomena and perceptual behavior.

References

Amari, S.: Dynamics of pattern formation in lateral-inhibition type neural fields. Biological Cybernetics 27, 77–87 (1977)

Baloch, A.A., Grossberg, S.: A neural model of high-level motion processing: Line motion and formotion dynamics. Vision Research 37, 3037–3059 (1997)

Bicho, E., Mallet, P., Schöner, G.: Target representation on an autonomous vehicle with low-level sensors. The International Journal of Robotics Research 19, 424–447 (2000)

Braitenberg, V.: Cortical architechtonics: General and areal. In: Brazier, M.A.B., Petsche, H. (eds.) IBRO Monograph Series, Architechtonics of the Cerebral Cortex, vol. 3, pp. 443–466. Raven Press, New York (1978)

Chey, J., Grossberg, S., Mingolla, E.: Neural dynamics of motion grouping: From aperture ambiguity to object speed and direction. Journal of the Optical Society of America A 14, 2570–2594 (1997)

Ditzinger, T., Haken, H.: The impact of fluctuations on the recognition of ambiguous patterns. Biological Cybernetics 63, 453–456 (1990)

Eastman, K., Hock, H.S.: Bistability in the perception of apparent motion: Effects of temporal asymmetry. Perception & Psychophysics 61, 1055–1065 (1999)

Eggert, J., van Hemmen, J.L.: Modeling neuronal assemblies: Theory and implementation. Neural Computation 13, 1923–1974 (2001)

Einstein, A.: On the movement of small particles suspended in a stationary liquid demanded by the molecular-kinetic theory of heat. Annalen der Physik 17, 540 (1905)

Felleman, D.J., Van Essen, D.C.: Distributed hierarchical processing in primate visual cortex. Cerebral Cortex 1, 1–47 (1991)

Francis, G., Grossberg, S.: Cortical dynamics of form and motion integration: Persistence, apparent motion, and illusory contours. Vision Research 36, 149–173 (1996)

Gibson, J.J.: The senses considered as perceptual systems. Houghton Mifflin, New York (1966)

Green, D.M., Swets, J.A.: Signal detection theory and psychophysics. Wiley, New York (1966)

Grossberg, S.: Contour enhancement, short term memory, and constancies in reverberating neural networks. Studies in Applied Mathematics 52, 213–257 (1973)

Grossberg, S., Mingolla, E.: Neural dynamics of motion perception: Direction fields, apertures, and resonant grouping. Psychological Review 92, 173–211 (1985)

Grossberg, S., Rudd, M.E.: Cortical dynamics of visual motion perception: Short-range and long-range apparent motion. Psychological Review 99, 78–121 (1992)

Hock, H.S., Balz, G.W.: Spatial scale dependent in-phase and anti-phase directional biases in the perception of self-organized motion patterns. Vision Research 34, 1843–1861 (1994)

Hock, H.S., Balz, G.W., Smollon, W.: The effect of attentional spread on self-organized motion perception. Vision Research 38, 3743–3758 (1998)

Hock, H.S., Brownlow, S., Taler, D.: The role of feedback in the formation of global motion patterns (in preparation)

Hock, H.S., Bukowski, L., Nichols, D.F., Huisman, A., Rivera, M.: Dynamical vs. judgmental comparison: Hysteresis effects in motion perception. Spatial Vision 18, 317–335 (2005)

Hock, H.S., Gilroy, L., Harnett, G.: Counter-changing luminance: A non-Fourier, nonattentional basis for the perception of single-element apparent motion. Journal of Experimental Psychology: Human Perception and Performance 28, 93–112 (2002)

Hock, H.S., Kelso, J.A.S., Schöner, G.: Bistability and hysteresis in the organization of apparent motion patterns. Journal of Experimental Psychology: Human Perception & Performance 19, 63–80 (1993)

Hock, H.S., Kogan, K., Espinoza, J.K.: Dynamic, state-dependent thresholds for the perception of single-element apparent motion: Bistability from local cooperativity. Perception & Psychophysics 59, 1077–1088 (1997)

Hock, H.S., Nichols, D.F., Espinoza, J.: When motion is not perceived: Evidence from adaptation and dynamical stability. Spatial Vision 17, 235–248 (2004)

Hock, H.S., Park, C., Schöner, G.: Self-organized pattern formation: Experimental dissection of motion detection and motion integration by variation of attentional spread. Vision Research 42, 991–1003 (2002)

Hock, H.S., Ploeger, A.: Linking dynamical decisions at different levels of description in motion pattern formation: Psychophysics. Perception & Psychophysics 68, 505–514 (2006)

Hock, H.S., Schöner, G., Giese, M.: The dynamical foundations of motion pattern formation: Stability, selective adaptation, and perceptual continuity. Perception & Psychophysics 65, 429–457 (2003)

Hock, H.S., Schöner, G., Hochstein, S.: Perceptual stability and the selective adaptation of perceived and unperceived motion. Vision Research 36, 3311–3323 (1996)

Hock, H.S., Schöner, G., Voss, A.: The influence of adaptation and stochastic fluctuations on spontaneous perceptual changes for bistable stimuli. Perception & Psychophysics 59, 509–522 (1997)

Jancke, D., Erlhagen, W., Dinse, H.R., Akhavan, A.C., Giese, M., Steinhage, A., Schöner, G.: Parametric population representation of retinal location: Neuronal interaction dynamics in cat primary visual cortex. Journal of Neuroscience 19, 9016–9028 (1999)

Johnson, J.S., Spencer, J.P., Schöner, G.: Moving to higher ground: The dynamic filed theory and the dynamics of visual cognition. New Ideas in Psychology 26, 227–251 (2008)

Leopold, D.A., Wilke, M., Maier, A., Logothetis, N.K.: Stable perception of visually ambiguous patterns. Nature Neuroscience 5, 605–609 (2002)

Nichols, D.F., Hock, H.S., Schöner, G.: Linking dynamical decisions at different levels of description in motion pattern formation: Computational simulations. Perception & Psychophysics 68, 515–533 (2006)

Poston, T., Stewart, I.: Nonlinear modeling of multistable perception. Behavioral Science 23, 318–334 (1978)

Schneegans, S., Schöner, G.: Dynamic field theory as a framework for understanding embodied cognition. In: Calvo, P., Gomila, T. (eds.) Handbook of Cognitive Science: An Embodied Approach, pp. 241–271 (2008)

Schöner, G.: Dynamical systems approaches to cognition. In: Sun, R. (ed.) Cambridge handbook of computational cognitive modeling, pp. 101–126. Cambridge University Press, Cambridge (2008)

Tanaka, K., Saito, H.-A.: Analysis of motion of the visual field by direction, expansion/contraction, and rotation cells clustered in the dorsal part of the medial superior temporal area of the macaque monkey. Journal of Neurophysiology 62, 643–656 (1989)

Trappenberg, T.P.: Fundamentals of computational neuroscience. Oxford University Press, Oxford (2002)

Treisman, A.: The binding problem. In: Squire, L.R., Kosslyn, S.M. (eds.) Findings and current opinion in cognitive neuroscience. The MIT Press, Cambridge (1998)

Ullman, S.: The interpretation of visual motion. MIT Press, Cambridge (1979)

Wilson, H.R., Cowan, J.D.: A mathematical theory of the functional dynamics of cortical and thalmic nervous tissue. Kybernetik 13, 55–80 (1973)

Wilson, H.R., Kim, J.: A model for motion coherence and transparency. Visual Neuroscience 11, 1205–1220 (1994)

Yuille, A.L., Kersten, D.: Vision as Bayesian inference: analysis by synthesis? Trends in Cognitive Science 10, 301–308 (2006)

Optical Illusions: Examples for Nonlinear Dynamics in Perception

Thomas Ditzinger

Abstract. Simple and intriguing examples for nonlinear dynamics in visual perception are presented by means of optical illusions. Well known visual effects such as the temporal perception of ambiguous figures, autostereograms, and moving patterns are presented and interpreted from the perspective of nonlinear dynamics. Furthermore new results on the interdependency between the perception of colour and motion are presented, including an explanation of the classic "Fluttering Hearts effect" and the new "Leaning Tower of Pisa effect" which is responsible for a perceptual shift of rotating coloured areas.

1 Ambiguous Figures

Ambiguous patterns are well known and fascinating examples for optical illusions.
Because of their reliability in the experimental procedure ambiguous figures are a popular tool in psychophysics. When test persons look at a figure as shown in Fig 1, they experience a temporal oscillation in perception, e.g. the three-dimensional cube becomes suddenly dynamic. The plane seen in the foreground suddenly appears to be lying in the background for some seconds, then in the foreground again, then again in the background, etc. The best known psychological hypothesis to explain these findings is that of saturation of perception, proposed by Köhler (1940). There is a general agreement among psychologists that the observed oscillations result from neuronal fatigue, inhibition or saturation.
This perceptual behaviour can be successfully mimicked by a nonlinear model of human perception of ambiguous patterns (Ditzinger, Haken 1989, 1990) which is in excellent agreement with the experimental findings. The model is a straightforward extension of a general algorithm for pattern recognition by machines known as synergetic computer (Haken 2004) on the one hand and of the saturation of attention parameters on the other. It simulates the recognition of ambiguous patterns by humans by a set of coupled differential equations which describe the formation of percepts by means of order parameters. The degree to which an indivdual pattern k is recognized by a subject is described by the order parameters

---

Thomas Ditzinger
Springer Verlag, Tiergartenstr, Heidelberg, Germany

ξ_k which in turn is determined by the saturation of the corresponding attention parameters λ_k. In our model (1-4) for the perception of ambiguous figures these attention parameters are subjected to a damping mechanism mimicking the effect of saturation of attention and synaptic connections with a,b, γ constant. In this way oscillations of perception arise quite naturally.

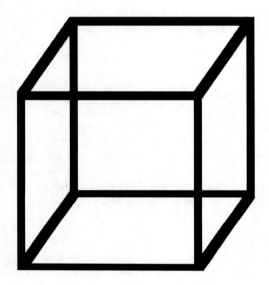

Fig. 1 The Necker cube (Necker 1840)

$$\dot{\xi}_1 = \xi_1[\lambda_1 - a\xi_1^2 - b\xi_2^2 + \alpha\xi_2^2(1 - \frac{2\xi_2^4}{(\xi_1^2 + \xi_2^2)^2})] \tag{1}$$

$$\dot{\xi}_2 = \xi_2[\lambda_2 - a\xi_2^2 - b\xi_1^2) + \alpha\xi_1^2(1 - \frac{2\xi_1^4}{(\xi_1^2 + \xi_2^2)^2})] \tag{2}$$

$$\dot{\lambda}_1 = \gamma(1 - \lambda_1 - \xi_1^2) + F_1(t) \tag{3}$$

$$\dot{\lambda}_2 = \gamma(1 - \lambda_2 - \xi_2^2) + F_2(t) \tag{4}$$

Our approach takes also ambiguous patterns with bias into account which leads to different periods of the attention paid to the one or the other interpretation of the pattern. The perceptual weight of a bias of a percept k is denoted by the bias parameter α. A typical simulation result can be seen in figure 2 including fluctuatios $F_k(t)$ of the attention parameters.

Optical Illusions: Examples for Nonlinear Dynamics in Perception

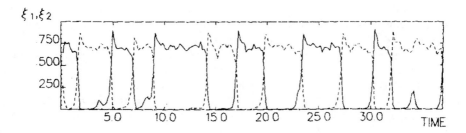

Fig. 2 Temporal behaviour of the order parameters ξ_1 and ξ_2. The typical oscillations of the perception can be clearly seen

Our model is in excellent agreement with the experimental findings and intuition. As can be seen in Fig. 2 when a certain alternative has been perceived, it remains stable over some time until suddenly the other alternative is perceived.

The two alternatives of perception are rarely of exactly equal strength. In most cases one is preferred against the other. Despite of that, if the bias of one alternative is not strong, a reversion occurs, but the reversion times of the individual components become different. The stronger the bias, the longer the reversion time.

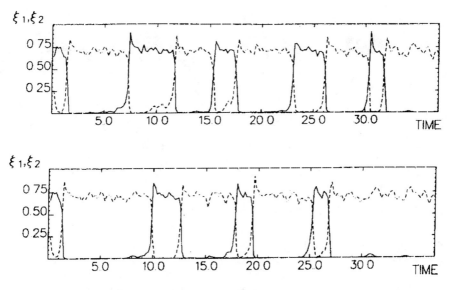

Fig. 3 Oscillations of perception between percepts with different bias (upper part: $\alpha = 0.064$, lower part: $\alpha = 0.128$.

Our model (1-4) shows exactly the same behaviour which can be seen in Fig. 3 for ambiguous figures with increased bias. In the upper part, the time evolution is shown for $\alpha = 0.064$, in the lower part for $\alpha = 0.128$. As can be seen the share of the alternative 2 (dashed line) is increased. With increasing bias also the sum of the individual reversion times increases in good agreement to the measurement.

With our model we can also show that the duration of the period as a sum of the reversion times becomes longer with increasing bias of one alternative of perception.

With our model we can also demonstrate the phenomenon of hysteresis in perception. If one considers the sequence of pictures of Fig. 4 from the left upper corner to the right lower corner and then in the opposite on the direction of view direction, one observes that the transition from the face of a man to a girl occurs at different points.

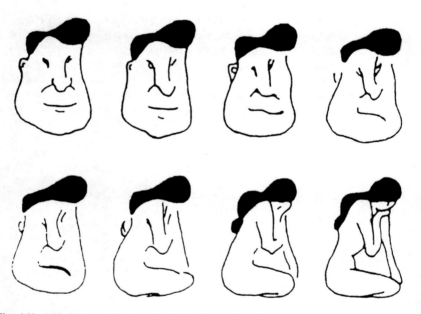

Fig. 4 Hysteresis in perception: the place of transition depends on the direction of view.

2 The Verbal Transformation Effect

While there are usually two alternatives reported in experiments with visual ambiguous figures, the number of reported alternative phonemic structures is typically much greater. The most general form of multistability in perception can be studied in hearing in the form of the verbal transformation effect. For the first time Richard Warren studies in 1958 the perception of listeners to repeatedly presented auditory stimuli such as syllables, words or phrases. He found that most of the subjects perceive the stimulus in variety of alternative forms.

Optical Illusions: Examples for Nonlinear Dynamics in Perception   183

In a recent experiment (Tuller et al. 1997) the syllable [ke] was reproduced 1000 times with a 500-ms silent interval between successive tokens and presented to subjects. All subjects reported hearing changes in the syllable presented (e.g., 'ke', 'ka', 'cat', 'kayah', 'chair'). Although subjects may report dozens of different alternative forms during the course of the experiment, listeners tend to cycle between only two forms at a time, not three or more.

It was found that the main organization of the perceptual transition is in pairs (Ditzinger et al. 1997). For example, when a single syllable was repeatedly presented, cycling between two phonemic forms was far more common than cycling among three or more forms. This means that the perceptual order for three alternatives 1,2,3 is perceived in paircoupled sequences such as 12121213131313232323 instead of e.g. circular sequences such as 123123123. It was shown that the paircoupled transforms of perception have a faster and more stable dynamics than the nonpaircoupled transforms.

As can be shown our model of multistable perception produces exactly the same temporal paircoupling behaviour as in the measurement. A straightforward generalization of our model (1-4) to more than 2 alternatives k leads to the full model

$$\dot{\xi}_k = \xi_k [\lambda_k - B \sum_{k=1}^{M} \xi_k^2 (1 - \alpha_{kk} (1 - \frac{2\xi_k^4}{(\xi_k^2 + \xi_k^2)^2})) + B\xi_k^2) - C\xi_1^2] \sum_{k=1}^{M} \xi_k^2]$$

$$\dot{\lambda}_k = \gamma(1 - \lambda - \xi_k^2) + F_k(t) \qquad k=1,2,...M \qquad (5)$$

There is very good agreement between the empirically observed properties of the verbal transformation effect and the properties detected by our full model. With our model it can be shown that the main organization of the perceptual transitions is into pairs as in the experiments. This pairwise coupling is pronounced in the frequency of switching to pair members, but not in their dwell times (the time spent perceiving a given phonemic form before switching to another form). As in the experiments the paircoupled transforms of perception have a faster and more stable dynamic than the non - paircoupled transforms.

3   3D Vision and Autostereograms

Recently, a very impressive method of encoding three-dimensional information in 2D pictures was designed in the form of computer-generated patterns of colored dots. At first glimpse, these so-called autostereograms appear as structured but meaningless patterns. After a certain period of observation, a 3D pattern emerges suddenly in an impressive way. In Fig. 5 an expample of an autostereogram is presented. After some time of observing the fiat, periodic random dot pattern, a striking phase transition in perception occurs, and 3D rabbits becomes visible.

Human 3D perception of autostereograms is only possible if the eyes see two slightly different images, although observing only one autostereogram. The left-image is any segment of the original autostereogram. The right image is a segment

Fig. 5 Example of an Autostereogram: Rabbits

of the same size but taken from a position horizontally shifted with respect to the left image. The amount of the shift is connected with the eye distance and the visual angle. In order to obtain a stereoscopic perception the convergence of the eyes has to be controlled and the sight has to go behind the paper plane. This means that the accommodation of the lenses to the paper plane and the convergence of the eyebeams (behind the paper plane) have to be decoupled. By means of the amount of the horizontal difference of the two pictures – the disparity- the corresponding image of the eyes are fused and a meaningful 3D perception becomes possible. This impressive transition between the initial state and the 3D perception state takes place in a very short time.

When humans concentrate long enough on an autostereogram, they can perceive a variety of different 3D by the vergence eye movement control aspects. Looking at an original autostereogram as in Fig. 5, the most probable and impressive aspect is the perception of two 3D rabbits. The eyes of the observer are initially uncrossed by a certain amount. The vergence eye movement stops when the horizontal shift from the left image to the right image is approximately one period length in dot sequence to the right. Other possible 3D aspects are induced by initially uncrossing the eyes even further, leading the vergence eye movement control to produce a horizontal shift of 2 or more period lengths. Thereby, the 3D background impression becomes deeper, and in the foreground, different 3D ghost images become apparent. In the case of a horizontal shift of 2 period lengths, six slighty different rabbits become visible: four in the background at the same depth and two, incomplete ones in the foreground. The shape of the incomplete rabbits

Optical Illusions: Examples for Nonlinear Dynamics in Perception

is the overlapping of the four background shapes. Some persons are able to perceive an inverse 3D picture, too. The background of the original uncrossed eye case becomes the foreground and vice versa. This is the result of initially crossing the eyes, producing a horizontal shift of the left image to the right image in the opposite direction as before.

All these characteristic properties of human perception of autostereograms can be simulated by a nonlinear model for stereo-vision (details see Reiman and Haken 1994) which was applied to the perception of autostereograms (Reimann et al. 1995). The model shows a fully satisfactory agreement with the multivalent perception experienced by humans. As in nature, in our model the phase transition between the initial state and the 3D perception state takes place in a very short time.

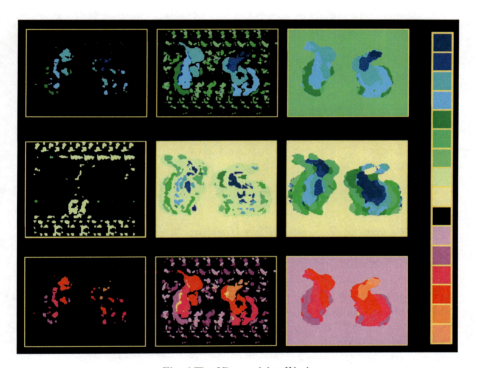

Fig. 6 The 3D - model at Work

The temporal development of the simulated depth maps of the model can be seen in Fig. 6.

In the upper part the temporal evolution of the depth map is shown for regular uncrossed 3D- viewing according to the linear scale given at the right hand side: blue indicates the foreground, orange the deepest background. In the first picture, the result for the first iteration steps is shown. Up to now, only a few 'conscious' parts of the 3D pattern are perceived by the system. In the course of time, the depth map suddenly completes itself in a self-organizing way. Order is created by competition and cooperation in a self-organizing process. The final state can be seen in the third picture, which clearly shows the two rabbits as perceived by humans.

Other possible 3D aspects are induced by initially uncrossing the eyes further, leading the vergence eye movement control to produce a horizontal shift of 2 or more period lengths. Thereby, the 3D background impression becomes deeper, and in the foreground, different 3D ghost images become apparent. As shown in the second row of Fig. 6 in the case of a horizontal shift of 2 period lengths, our model perceives six different rabbits: four in the background at the same depth and two more, incomplete ones in the foreground. The shape of the incomplete rabbits is the overlapping of the two background shapes. This is in full accordance to human perception.

The case of inverse 3D picture recognition can be simulated in full amount, too. This can be seen in the bottom part of Fig. 6. After some initial time steps suddenly a clear percept appears in inverted depths. In the third figure of the third row you can see the orange ear of the right hand rabbit indicating the deepest level of the depth map, while the purple plain in this inverted view indicates the perceived foreground.

Our nonlinear model has proven to be fully capable to simulate the properties of human 3D vision processes in the perception of autostereograms. As in human perception our model shows multistability in autostereogram perception according to the number and 3D- positioning of perceived rabbits.

4 Perception of Motion, Colour and Brightness

4.1 The Leaning Tower of Pisa Illusion

A complete nonlinear dynamic model of human perception should also be able to simulate the perception of motion, colour and brightness. In this paragraph we focus on a new, striking and easily accessible optical illusion mediated by interactions of colour, brightness, form and motion perception – the Leaning tower of Pisa (LTOP) illusion (Ditzinger et al 2000). In some circumstances the perception of orientation of coloured forms is radically altered by rotary movement. This illusion can be easily demonstrated with a common record turntable. By using optimized colour and brightness combinations between foreground and background an illusory tilt of 8 degrees and more can be observed.

In Figure 7 a coloured Leaning tower of Pisa is visible with its original 5,5 degrees tilt. At speeds obtainable form a record turntable (33-78 rpm) the tilt disappears! The LTOP effect is also visible in the orientation of the yellow bar symbolising the horizon which upon turntable rotation appears to be parallel to the bottom of the cyan picture frame. Further experiments show that this phenomenon is dependent on rotation rate and on the used combination of colour and luminance.

Optical Illusions: Examples for Nonlinear Dynamics in Perception 187

Fig. 7 The Leaning Tower of Pisa Illusion. The original tilt of 5,5 degrees disappears when rotated on a regular turntable player and the tower appears to be upright.

This can be impressively demonstrated by means of a pattern (see figure 8) with a set of parallel blue lines on a red background. Additionally there is a green and a yellow line, tilted 6° clockwise from the set of parallel blue lines. Upon rotation the green line appears parallel to the blue lines, but the tilted yellow line appears still tilted. This shows that the colour combination red/green produce a strong LTOP effect, while the combinations red/blue or red/yellow produce no effect or only a weak one. The effect is dependent on the direction of movement. If e.g. Fig. 7 or 8 are rotated in the opposite way it can be seen that the orientation of the targets (tower in Fig. 7, green line in Fig. 8) appears exaggerated by movement.

The LTOP illusion appears to be in good agreement with the venerable fluttering heart illusion (Wheatstone 1844, Brewster 1844, von Helmholtz 1867). If a coloured target attached to a differently coloured background is shaken in dim

Fig. 8 The yellow and the green line is tilted 6° clockwise from a set of parallel blue lines. When spun on a turntable player the green line appears to be parallel to the blue lines but not the yellow.

illumination and viewed in peripheral vision, then for some colour combinations the target appears to become detached from the background and to lead or lag the background motion with a characteristic delay. We found that the interacting combinations of colour and brightness are the same as in the LTOP illusion. The fluttering heart illusion is less intense and stable than the LTOP illusion though And it is restricted to dim illumination and peripheral vision, which is not the case for the LTOP illusion. A possible explanation for the perceptual strength of the LTOP illusion is the different geometry of target movements. While the movement in the fluttering heart illusion is a one-dimensional oscillation with variable velocity and speed, the LTOP movement is a uniform circular motion (in two dimensions) with constant speed. This constant speed produces constant temporal viewing conditions - in contrast to the fluttering heart illusion.

Optical Illusions: Examples for Nonlinear Dynamics in Perception   189

As of now we know of no complete explanation and modelling of the fluttering heart illusion nor the LTOP phenomenon but hypothesize that the effect is connected with the different temporal delays in the perception of different colours and their luminances.

A new finding on the way to an explanation is that each colour not only has an interacting motion partner in colour but also in greyscale. This can be seen in the upper part of Fig. 9 showing 11 areas with different greyscale values on a magenta background (greyscale =0.0, 0.1, 0.2.. 1.0). Dependent on the printer conditions one of the grey areas appears to detach from ground when shaken and move with a delay. The same result occurs with the LTOP experiment. This means that the detaching phenomenon is not only caused by interacting ingenious pairs of colours but by a more general process connected to greyscale values.

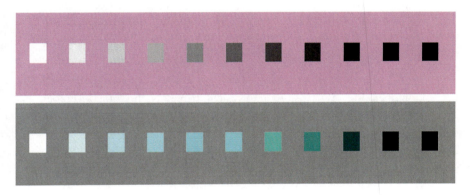

Fig. 9 Detecting coherent colours

In the lower part of Fig 9 it can be seen that also the opposite behaviour occurs: for each greyscale background an interacting coloured foreground can be found in each colour tone. This is demonstrated for a background with the greyscale value 0.4 and 11 cyan areas with different brightness values (C=0., 0.1, 0.2..). At least one detaching area can be perceived by each subject with the fluttering heart illusion as well as the LTOP phenomenon.

In each case a detachment effect appears for a special value of C. As measurements with a spotmeter show the detaching areas have comparable brightness values as their background! This means that there is not only a special pair of interacting colours but that every colour can shake with every other if just the luminance is right! We have tested this hypothesis with our printer for various colour combinations – for all of them the existence of the transitivity law was fullfilled. The detection procedure can be demonstrated by the example of Fig 9. For our printer in the upper part the area with 0.4 grey value interacted the most with the magenta background. Using this background in the lower part we found for our conditions and cyan foreground that the 50% luminance area in the middle interacted the most. And we finally saw, that this colour also fully interacts with the magenta background from the upper part of Fig.9!

Due to this circular transitivity it is reasonable to hyothesize, that the areas appear to detach from the ground if they have exactly the same perception times. If the perception times are of the same size a resonance effect takes place. This resonance effect is not possible for different perception times, because they produce a ambiguity in spatial perception of the areas: on some locations the areas would appear to intersect and on some locations none of the areas would be unambiguous. Therefore it looks like our cognition system overrules this inconsistent visual input in the form of a new Gestalt law: the uniformity of motion. This may be done by using haptic information and experience to produce a well-defined perceipt. This is not the case for exactly the same perception times: no ambiguity appears, the different areas are able to move with the same time delay.

It looks like that there are important consequences to well known optical illusions using the interaction of coherent colour pairs. We saw that these colours reduce the illusion of the stereokinetic effect/rotoreliefs drastically. They also reduce the effect of the Ouchi apparent motion illusion and the wakes and spokes illusion.

Impacts to the Benhams disks are under consideration.

These new findings should help to build a complete nonlinear model of human perception including 3D- vision, multivalent perceipts, brightness, colour and motion in the near future.

References

Brewster, D.: On the same subject, Report of the British Association for the Advancement of Science, Transactions Section 10 (1844)

Ditzinger, T., Haken, H.: Oscillations in the perception of ambiguous patterns. Biological Cybernetics 61, 279–287 (1989)

Ditzinger, T., Haken, H.: The impact of fluctuations on the recognition of am-biguous patterns. Biological Cybernetics 63, 453–456 (1990)

Ditzinger, T., Tuller, B., Kelso, J.A.S.: Temporal patterning in an auditory illusion: the verbal transformation effect. Biological Cybernetics 77, 23–30 (1997)

Ditzinger, T., Tuller, B., Haken, H., Kelso, J.A.S.: A synergetic model for the verbal transformation effect (1997b)

Ditzinger, T., Billock, V.A., Kelso, J.A.S., Holtz, J.: The leaning Tower of Pisa Effect: An illusion mediated by colour, brightness, and motion. Perception 29, 1269–1272 (2000)

Haken, H.: Synergetic Computers and Cognition, 2nd edn. Springer, Heidelberg (2004)

von Helmholtz, H.: Handbuch der Physiologischen Optik, Voss, Hamburg (1867)

Koehler, W.: Dynamics in psychology. Liveright, New York (1940)

Necker, L.A.: Observations on some remarcable phenomenon which occurs on viewing a figure of a crystal or geometrical solid. London and Edinburgh Philosophical Magazine and Journal of Science 3, 329–337 (1832)

Reimann, D., Haken, H.: Stereo-vision by selforganization. Biological Cybernetics 71, 17–26 (1994)

Reimann, D., et al.: Vergence Eye Movement Control and Multivalent Perception of Autostereograms. Biological Cybernetics 73, 123–128 (1995)

Tuller, B., Ding, M., Kelso, J.A.S.: Fractal timing of phonemic transforms. Perception 26, 913–928 (1997)

Warren, R., Gregory, R.: An auditory analogue of the visual reversible figure. American Journal of Psychology 71, 612–613 (1958)

Wheatstone, C.: On the singular effect of the juxtaposition of certain colours under particular circumstances. Report of the British Association for the Ad-vancement of Science, Transactions Section 10 (1844)

A Dynamical Systems Approach to Musical Tonality

Edward W. Large

Abstract. Music is a form of communication that relies on highly structured temporal sequences comparable in complexity to language. Music is found among all human cultures, and musical languages vary across cultures with learning. Tonality – a set of stability and attraction relationships perceived among musical frequencies – is a universal feature of music, found in virtually every musical culture. In this chapter, a new theory of central auditory processing and development is proposed, and its implications for tonal cognition and perception are explored. A simple model is put forward, based on knowledge of auditory organization and general neurodynamic principles. The model is simplified as compared to the organization and dynamics of the real auditory system, nevertheless it makes realistic predictions about neurodynamics. The analysis predicts the existence of natural resonances, the potential for tonal language learning, the perceptual categorization of intervals, and most importantly, relative stability and attraction relationships among musical tones. This approach suggests that high-level music cognition and perception may arise from the interaction of acoustic signals with the dynamics of the auditory system. Musical universals are predicted by intrinsic neurodynamics that provide a direct link to neurophysiology, and Hebbian synaptic modification could explain how different tonal languages are established.

1 Introduction

The music of almost every instrumental culture is tonal. In tonal music, one specific tone, called the tonic, provides a focus around which other tones are organized. Musical melodies typically involve discrete tones, organized in

---

Edward W. Large
Center for Complex Systems and Brain Sciences
Florida Atlantic University 777 Glades Road, Bldg 12, Rm 327,
Boca Raton, FL 33431
Ph.: 561.297.0106
Fax: 561.297.3634
e-mail: large@ccs.fau.edu

archetypal patterns that are characteristic of musical genres, styles, and cultures. These patterns may be related to a scale, an ordered collection of all the tones used in a given melody, summarizing the frequency ratios that govern the intervals between tones in a melody. The Western tuning system defines an inclusive scale, the chromatic scale, which divides the octave into 12 steps, called semitones. A subset of these 12 tones, called a diatonic scale, is typically used to create a melody. In modern Western tonality, the main diatonic scales are major and minor. When a melody is in a key, say C major, the notes of the C major scale are used to create that melody. One feature that the melodies of most musical systems share is that they give rise to tonal percepts. The tonic – C, in the key of C major – is said to be the most stable tone in that key. Stability means that the tone is perceived as a point of repose. For example, a melody in the key of C major will almost always end on the tonic, C. Among the other tones of the scale, there is a hierarchy of relative stability, such that some tones are perceived as more stable than others. Less stable tones provide points of dissonance or tension, more stable tones provide points of consonance or relaxation. Finally, less stable tones are heard relative to more stable ones, such that more stable tones are said to attract the less stable tones.

What processes and network architectures in the nervous system could give rise to such perceptions in music? This chapter argues that nonlinear neural resonance underlies the perception of tonality. Universal properties of nonlinear resonance predict the perception of stability and attraction in tonal music as well as preferences for small integer ratios and perceptual categorization. Such principles could provide a set of innate constraints that shape human musical behavior and enable the acquisition of musical knowledge.

2 Tonality

The oldest theory of musical consonance is that perceptions of consonance and dissonance are governed by ratios of whole numbers. Pythagoras is thought to have first articulated the principle that intervals of small integer ratios are pleasing because they are mathematically pure (Burns, 1999). He used this principle to explain the musical scale that was in use in the West at the time, and Pythagoras and his successors proposed small-integer-ratio systems for tuning musical instruments, such as Just Intonation (JI) (see Table 1). Because transposition on fixed tuning instruments, like the piano, is problematic for JI, modern Western equal temperament (ET), divides the octave into 12 intervals that are precisely equal on a log scale. ET approximates JI, and transposition in ET is perfect, because the frequency ratio of each interval is invariant. However, aside from octaves the intervals are not small integer ratios, they are irrational. The fact that equal tempered intervals sound approximately as consonant as neighboring small-integer-ratio intervals is generally considered *prima facie* evidence against the Pythagorean theory of musical consonance.

Helmholtz (1863) hypothesized that the dissonance of a pair of simultaneously sounding complex tones was due to the interference of its pure tone components, explaining dissonance as an unpleasant sensation of roughness produced by the beating of sinusoids. This phenomenon, called sensory dissonance, is heard when tones interact within an auditory critical band (Plomp & Levelt, 1965), and the interaction of pure tone components correctly predicts ratings of consonance for pairs of complex tones (Kameoka & Kuriyagawa, 1969). However, sensory consonance does not fully explain the perception of musical consonance. For one thing, the sensory dissonance phenomenon applies to isolated clusters of simultaneously sounded tones, whereas musical consonance and dissonance are intrinsically dynamic: " ... a dissonance is that which requires resolution to a consonance" (Dowling, 1978).

Table 1 Tuning systems: Tone frequencies are chosen to divide the octave into (approximately) equal steps. Just intonation uses small integer ratios; equal temperament provides an approximation. The major scale (white notes) is one diatonic subset used to create melodies.

	Just Intonation	JI Decimal Equivalents	Equal Temperament
	1:1	1.000	1.000
	16:15	1.067	1.060
	9:8	1.125	1.123
	6:5	1.200	1.189
	5:4	1.250	1.260
	4:3	1.333	1.335
	45:32	1.406	1.414
	3:2	1.500	1.498
	8:5	1.600	1.587
	5:3	1.667	1.682
	7:4	1.750	1.782
	15:8	1.875	1.888
	2:1	2.000	2.000

In tonal music, some tones are perceived as more stable than others (Krumhansl, 1990; Lerdahl, 2001). More stable tones function as points of relative rest, whereas less stable tones tend to resolve to more stable ones. Theoretically, the stability of each pitch class relative to the other pitch classes is often described as a hierarchy (e.g., Lerdahl, 2001; see Figure 1A). Tones that are more stable occupy higher levels in the hierarchy; tones on a lower level are heard in relation to tones on the adjacent higher level. Krumhansl and Kessler (Krumhansl & Kessler, 1982) asked listeners to rate how well individual pitches fit within a tonal context. Such experiments provide profiles that quantify the stability of each tone within a musical key (see Figure 4, below). When applied to Western tonal contexts, the measured hierarchies are found to be consistent with music-theoretic accounts (Krumhansl, 1990; Krumhansl & Kessler, 1982). Moreover, stability measures correlate well with empirical frequencies of occurrence of tones in tonal

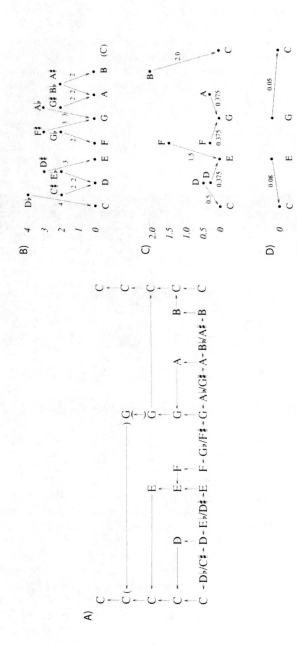

Fig. 1 Music theoretic depictions of tonal stability and attraction. A) The basic pitch class space for a tonic chord in the key of C Major. Attractions to superordinate neighbors are shown separately for each level of the basic space: chromatic (B), diatonic (C) and triadic (D). Arrows indicate goals and sizes of attractions. Reproduced from Lerdhal, 2001.

songs (e.g. Knopoff & Hutchinson, 1983). The strong expectancy for less stable tones to resolve to more stable ones is called *attraction*. Some theorists have described tonal attraction by analogy to physical forces, such as gravity and inertia (Larson, 2004); others link it to the resolution of musical dissonance (Bharucha, 1984). Transition probabilities in databases of folksongs (e.g., Eerola & Toiviainen, 2004), as well as listener's expectations about the completion of musical sequences (Bharucha & Stoeckig, 1986; Cuddy & Lunney, 1995; Larson, 2004), all show a strong influence of tonal attraction. Lerdahl's tonal pitch space summarizes such findings, quantifying net resultant attraction of pitches at one level of a stability hierarchy toward pitches at the next level of stability (Lerdahl, 2001) as shown in Figure 1, Panels B, C and D.

The tuning systems of the world's largest musical cultures, Western, Chinese, Indian, and Arab-Persian, are based on small integer-ratio relationships (Burns 1999). However, each tuning system is different, and this has led to the notion that frequency relationships do not matter in high level music cognition; rather, auditory transduction of musical tones results in abstract symbols, as in language. If this were true, stability and attraction relationships would have to be learned solely based on the frequency-of-occurrence statistics of tonal music (e.g. Krumhansl, 1990; Tillmann, Bharucha, & Bigand, 2000). However, this general approach does not explain why the statistics of tonal music would develop as they have; it assumes that statistical properties are given *a priori*. It also does not explain the significance of different tuning systems; it would make the same predictions given two different sets of tone frequencies with the same statistical relationships. The main hypothesis of this chapter is that nonlinear resonance in the central nervous system underlies the perception of tonality. Universal characteristics of nonlinear resonance predict that the perception of stability, tonal attraction, preference for small integer ratios, and perceptual categorization, are intrinsic to the dynamics of neural processing. Hebbian learning, which requires only passive exposure, accommodates the heterogeneity of tonal systems while the physics of nonlinear resonance predicts constraints on what can be learned.

3 A Dynamical Systems Approach

Interaction of excitatory and inhibitory neurons, illustrated schematically in Figure 2A, can give rise to neural oscillation. The current discussion follows the analysis of neural oscillation by Hoppenstadt and Izhikevich (Hoppenstadt & Izhikevich, 1996a, 1996b, 1997), extending the analysis to gradient frequency networks of neural oscillators driven by acoustic stimuli (Large, Almonte, & Velasco, 2010). Neural oscillation can be modeled theoretically using Equation 1 (Wilson & Cowan, 1973), which consists of two variables, describing the activity of excitatory (x) and inhibitory (y) neural populations:

$$\dot{x} = -x + S(\rho_x + ax - by)$$
$$\dot{y} = -y + S(\rho_y + cx - dy)$$
(1)

The overdot represents differentiation with respect to time, $\dot{x} = dx/dt$, and S is a sigmoid function. ρ_x and ρ_y are bifurcation parameters. Gradient frequency neural oscillator networks – under the influence of external input – can be written as

$$\tau_i \dot{x}_i = f_i(x_i, \lambda) + \varepsilon g_i(x, \lambda, \rho, \varepsilon) \tag{2}$$

where x is the 2-dimensional state vector for an oscillator, λ is a vector of parameters, ρ represents external input, and ε is the strength of the coupling nonlinearity (Hoppenstadt & Izhikevich, 1996a). The parameters are chosen such that $\tau_i = 1/f_i$, where f_i is natural frequency in Hz. The behavior of an oscillator under the influence of external input (Figure 2A) can be understood in detail by rewriting it in normal form (Wiggins, 1990; see Equation 3). The analysis involves a coordinate transformation, followed by Taylor expansion of the nonlinearities, truncating at some point to eliminate high order terms (abbreviated as h.o.t. in Equation 3). This results in z, a new, complex valued state variable, resulting from the coordinate transformation, and complex-valued parameters a and b ($a = \alpha + i\omega$, where ω is the radian frequency, $2\pi f$, and $b = \beta + i\delta$), which can be related to the parameters of the original system. $x(t)$ represents external input, either from another oscillator or from an acoustic signal.

$$\dot{z} = z(a + b|z|^2) + c\, x(t) + \text{h.o.t.} \tag{3}$$

The parameter, c, determines input connectivity and here we assume it to be real, although in general it could be complex (Hoppensteadt & Izhikevich, 1996a). The external stimulus $x(t) = A(t)e^{i\theta(t)}$ has frequency ω_0.

The behavior of this system can be understood by transforming it to polar coordinates using the relation $z = r(t)e^{i\phi(t)}$. This allows the independent study of amplitude, r, and phase, ϕ, dynamics of the oscillator.

$$\begin{aligned}\dot{r} &= r(\alpha + \beta r^2) + cA\cos(\theta - \phi) + \text{h.o.t.} \\ \dot{\phi} &= \omega + \delta r^2 + c\frac{A}{r}\sin(\theta - \phi) + \text{h.o.t.}\end{aligned} \tag{4}$$

Interactions between oscillators of different frequencies are found in the higher order terms. To understand these interactions, we expand out the higher order terms, keeping in mind that any resonant relationship among oscillator frequencies of the form

$$m_1\omega_1 + \ldots + m_n\omega_n + m_R\omega_R = 0 \tag{5}$$

is a resonance among eigenvalues of the uncoupled system and thus cannot be eliminated (Hoppenstadt & Izhikevich, 1997). This includes harmonics, subharmonics, and summation and difference tones of various orders. For example, for a pair of oscillators with $\omega_1 = 2\omega_2$

$$\begin{aligned}\dot{z}_1 &= z_1(a + b|z_1|^2) + \sqrt{\varepsilon}c_{12}z_2^2 + O(\varepsilon) \\ \dot{z}_2 &= z_2(a + b|z_2|^2) + \sqrt{\varepsilon}c_{21}z_1\bar{z}_2 + O(\varepsilon)\end{aligned} \tag{6}$$

A Dynamical Systems Approach to Musical Tonality

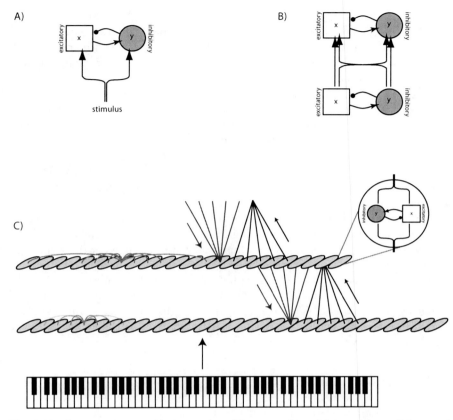

Fig. 2 Nonlinear resonance. A) A neural oscillator consists of interacting excitatory and inhibitory neural populations. B) Four synapses are possible from one neural oscillator to another. Changes in synaptic efficacy affect both the strength and the phase of oscillators' interaction and can be modified via Hebbian learning. C) A multi-layered, gradient frequency nonlinear oscillator network for responding to auditory stimulation.

and for a pair of oscillators with $\omega_1 = 3\omega_2$

$$\dot{z}_1 = z_1(a_1 + b_1|z_1|^2 + \varepsilon d_1|z_1|^4) + \varepsilon c_{12} z_2^3 + O(\varepsilon\sqrt{\varepsilon})$$
$$\dot{z}_2 = z_2(a_3 + b_2|z_2|^2 + \varepsilon d_2|z_2|^4) + \varepsilon c_{21} z_1^2 \bar{z}_2 + O(\varepsilon\sqrt{\varepsilon})$$
(7)

Carrying the analysis out further leads to a canonical model for gradient-frequency networks of nonlinear neural oscillators (Large, et al., 2010):

$$\tau_i \dot{z}_i = z_i(a + b_1|z_i|^2 + \varepsilon b_2|z_i|^4 + ...) + (x + \sqrt{\varepsilon}x^2 + \varepsilon x^3 + ...) \cdot (1 + \sqrt{\varepsilon}\bar{z}_i + \varepsilon \bar{z}_i^2 + ...)$$
(8)

The simulations reported in the paper were based on numerical solution of this differential equation (see Large, Almonte & Velasco, 2010 for further details).

Next, consider analysis of sound by the auditory system. Acoustic signals stimulate the cochlea, which performs a nonlinear time-frequency transformation (e.g., Camalet, Duke, Julicher, & Prost, 1999; Ruggero, 1992). Central auditory networks in cochlear nucleus, inferior colliculus, thalamus, and primary auditory cortex phase-lock action potentials to both sinusoidal and amplitude modulated (AM) signal features, further transforming the stimulus (Langner, 1992). Phase-locking deteriorates at higher-frequencies as the auditory pathway is ascended. The role of neural inhibition in the central auditory system is not yet fully understood. However, phase-locked inhibition exists in many auditory nuclei and plays a role in the temporal properties of neural responses (Grothe, 2003; Grothe & Klump, 2000) that could be consistent with nonlinear resonance. A simple model consistent with the known facts and the hypothesis of nonlinear resonance in the auditory system is illustrated in Figure 2C. It is based on networks of neural oscillators, in which each is tuned to a distinct natural frequency, or *eigenfrequency*, following a frequency gradient, similar in concept to a bank of bandpass filters. Within this framework, the input, x, to a gradient-frequency network of neural oscillators, would consist of afferent, internal and efferent input. For a network responding directly to an auditory stimulus, the afferent input would correspond to a sound. Despite the fact that the physiology of neural oscillation of oscillators can vary greatly, all nonlinear oscillators share many universal properties, providing certain degrees of freedom and also significant constraints, discussed next.

4 Predicting Tonality

Nonlinear resonance. Nonlinear oscillators possess a filtering behavior, responding maximally to stimuli near their own eigenfrequency. This is sometimes referred to as frequency selective amplification, due to extreme sensitivity to low amplitude stimuli. The first simulation (Figure 3) modeled frequency transformation of a sinusoidal stimulus by a single layer network of critical nonlinear oscillators (Equation 8), to demonstrate some basic properties. For this simulation the parameter values $\alpha = 0$; $\omega = 2\pi$; $\beta_1 = \beta_2 = \beta_n = -1$; $\delta_1 = 1$; $\varepsilon = 1.0$ were used, and $\tau_i = 1/f_i$, where f_i is the natural frequency of each oscillator in Hz. All other parameters were set to zero. The frequencies of the network were distributed along a logarithmic frequency gradient, with 120 oscillators per octave, spanning four octaves. The choice of $\alpha = 0$; $\beta_n < 0$ makes this a critical nonlinear oscillator, network similar to models that have been proposed for cochlear hair cell responses (Camalet, et al., 1999). No internal network connectivity was used in this simulation.

Figure 3 shows how a nonlinear oscillator bank responds as stimulus intensity varies. At low levels, high frequency selectivity is achieved. As stimulus amplitude increases, frequency selectivity deteriorates due to nonlinear excitation. As a nonlinear oscillator responds to a stimulus near its eigenfrequency, frequency entrains to that of the stimulating waveform, such that instantaneous frequency

A Dynamical Systems Approach to Musical Tonality

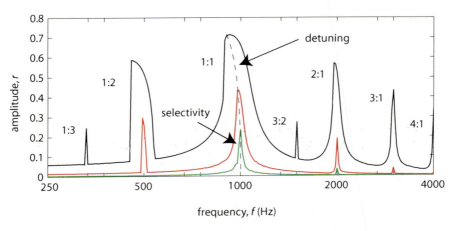

Fig. 3 Response amplitudes, r, of a gradient frequency nonlinear oscillator array (frequencies $250 \leq f \leq 4000\ Hz$) to a sinusoid (frequency $f_0 = 1000\ Hz$) at three different stimulus amplitudes.

comes to match stimulus frequency. A nonlinear oscillator array also responds at frequencies that are not physically present in the acoustic stimulus. At low stimulus intensities, higher-order resonances are small; they increase with increasing stimulus intensity. The strongest response is observed at the stimulus frequency, and additional responses are observed at harmonics and subharmonics of the sinusoidal stimulus. The second sub- and super-harmonics (1:2 and 2:1) are the strongest resonances, predicting the universality of the octave. Additionally, the response frequency of the oscillator depends on the amplitude of the resonance, i.e., frequency changes as amplitude increases. Such *frequency detuning* can be seen in Figure 3 as a bend in the resonance curve as stimulus intensity, and therefore, response amplitude increases. Frequency detuning could predict systematic departures from ET (and JI), which are commonly observed in category identification experiments (Burns, 1999), including octave stretch, as discussed below.

Natural resonances. For multi-frequency stimulation, the response of an oscillator network may include harmonics, subharmonics, integer ratios, and summation and difference tones, some of which are illustrated in Figure 3. To explore the natural resonances in a gradient frequency network a bifurcation analysis was used. Analysis of the higher-order resonances was based on the phase equations:

$$\dot{\phi}_1 = \omega_1 + c_{12}\, \varepsilon^{(k+m-2)/2} \sin(k\phi_2 - m\phi_1)$$
$$\dot{\phi}_2 = \omega_2 + c_{21}\, \varepsilon^{(k+m-2)/2} \sin(m\phi_1 - k\phi_2) \qquad (9)$$

where the frequency ratio is $k{:}m$ and the effects of amplitude are neglected. Here $c_{ij}\, \varepsilon^{(k+m-2)/2}$ is the strength, or relative stability, of the $k{:}m$ resonance, where c_{ij}

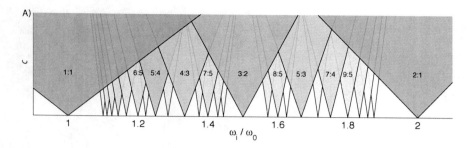

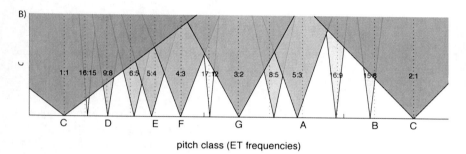

Fig. 4 Resonance regions. A) Bifurcation diagram showing natural resonances in a gradient frequency nonlinear oscillator array as a function of connection strength, C, and frequency ratio ω_i/ω_0. An infinite number of resonances are possible on this interval; the analysis considered the unison (1:1), the octave (2:1) and the twenty-five most stable resonances in between. B) Bifurcation diagram for a nonlinear oscillator network with internal connectivity reflecting an equal tempered chromatic scale. Internal connectivity can be learned via a Hebbian rule given passive exposure to melodies. Resonance regions whose center frequencies match ET ratios closely enough are predicted to be learned.

is coupling strength, a parameter that would be learned, and ε is the degree of nonlinearity in the coupling. This bifurcation analysis (Figure 4) assumes $\varepsilon=1$ (maximal nonlinearity) and plots resonance regions as a function of coupling strength on the vertical axis and relative frequency, ω_i/ω_0, on the horizontal axis. The phase equations (9) were used to derive the boundaries of the resonance regions, or Arnold tongues according to

$$\frac{k}{m} \pm c\frac{m+k}{mk}.$$

The analysis varied oscillator frequency and coupling strength, assuming equal stimulation to each oscillator at a fixed frequency (the tonic, ω_0). The result depicts the long-term stability of various resonances in the network, displayed as a bifurcation diagram called Arnold tongues (Figure 4A). It predicts how different pools of neural oscillators will respond by showing the boundaries of resonance neighborhoods as a function of coupling strength and frequency ratio. The oscillators in each resonance region frequency-lock at a specific ratio with the stimulus.

An infinite number of resonances exist on the interval between 1:1 and 2:1 and are found at integer ratios; smaller integer ratios are more stable and therefore more likely to be observed in the limit. This analysis displayed the 25 largest resonance regions on this interval, to provide a picture of the "natural" resonances in such a network. From the point of view of a gradient-frequency oscillator network, the Arnold tongues can be thought of as displaying the resonance for each oscillator in the network interacting with an oscillator outside the network (for example, an oscillator providing afferent stimulation) whose frequency corresponds to the tonic (1:1). The analysis considers pairwise interactions only; it neglects interactions between the oscillators within the gradient frequency network. Thus, it provides a somewhat simplified picture of network behavior, but one that is highly informative.

Nonlinear resonance predicts a generalized preference for small integer ratios. This prediction does not correspond to any specific musical scale; rather, natural resonances predict constraints on which frequency relationships can be learned. When stimulus frequency does not form a precise integer ratio with the eigenfrequency of an oscillator, resonance is still possible, provided that coupling is strong enough. Resonances affect not only oscillators with precise integer ratios; they also establish patterns of resonant neighborhoods.

Learning. Hebbian learning provides a theoretical basis for the acquisition of tonality relationships. Connections between oscillators can be learned via a Hebbian rule (Hoppensteadt & Izhikevich, 1996b), providing a mechanism for synaptic plasticity wherein the repeated and persistent co-activation of a presynaptic cell and a postsynaptic cell lead to an increase in synaptic efficacy between them. Between two neural oscillators four synapses are possible (Figure 2B, above), providing both a strength and a natural phase for the connection between neural oscillators (Hoppensteadt & Izhikevich, 1996a). Hebbian learning rules have been proposed for neural oscillators and the single-frequency case has been studied in some detail (Hoppensteadt & Izhikevich, 1996b). For the single-frequency case, the Hebbian learning rule can be written as follows:

$$\dot{c}_{ij} = -\delta c_{ij} + k_{ij} z_i \bar{z}_j \qquad (10)$$

This model can learn both amplitude and phase information for two oscillators with a frequency ratio near 1:1. In the current study, learning of higher-order resonances is also of interest. The following generalization of the above learning rule enables the study of learning for higher order resonant relationships

$$\dot{c}_{ij} = -\delta c_{ij} + k_{ij}(z_i + \sqrt{\varepsilon}z_i^2 + \varepsilon z_i^3 + ...) \cdot (\bar{z}_j + \sqrt{\varepsilon}\bar{z}_j^2 + \varepsilon \bar{z}_j^3 + ...). \qquad (11)$$

Coupling strength, c_{ij}, is the parameter that would be altered by learning. Due to computational complexity, extensive simulation of learning on melodies has not yet been carried out. However, analysis of multi-frequency Hebbian learning

(Eq. 11) shows that connections to near-resonant frequencies, such as integer ratios, can be learned in a gradient frequency network. Assuming stimulation with melodies using ET tone frequencies, connections would be learned between an oscillator at the frequency of the tonic (1:1) and the most stable resonances that approximate the ET tone frequencies closely. Thus, a second bifurcation analysis was performed, in which the k and m parameters were chosen as the largest resonance region (smallest integer ratio) that approximated the ET tone frequency to within 1%[1]. The result of this analysis is shown in Figure 4B. Learning would likely result in different coupling strengths for each resonance, therefore analysis shows each resonance region for a range of coupling strengths.

This analysis predicts that, as Western melodies are heard, the network would learn the most stable attractors whose center frequencies closely approximate the ET chromatic frequencies. Hebbian synaptic modification would effectively prune some resonances, while retaining others. The resulting resonances reflect the chromatic scale as shown in Figure 4B. In principle a similar learning analysis could be performed for any tuning system, such as gamelan, whose frequency ratios differ significantly from 12-tone ET.

Perceptual Categorization. In Figure 4B, the center frequencies of the resonances do not precisely match ET frequencies; however, as connection strength increases, larger regions of the network resonate, emanating from integer ratios, and encompassing ET ratios. Such regions predict perceptual categorization of musical intervals. Perceptual categorization and discrimination experiments reveal that musicians show categorical perception of melodic intervals (Burns & Campbell, 1994), and although such experiments are more difficult with non-musicians, Smith et al. (Smith, Nelson, Grohskopf, & Appleton, 1994) demonstrated that nonmusicians also perceive pitch categories. Dependence of frequency on amplitude further predicts that perceptual categories might not be precisely centered on integer ratios. In interval identification experiments, mean frequency deviates systematically from ET, although not always in the direction predicted by JI. Musicians prefer flatter small intervals and sharper large intervals (Burns, 1999). In fact, in many tuning systems (including the piano) octaves are stretched, i.e., tuned slightly larger than 2:1. In performance on instruments without fixed tuning (e.g., the violin, or the human voice), mean frequency also deviates systematically, similarly to perceptual categorizations (Loosen, 1993). More importantly, frequency variability is quite large in music performance, even during the "steady state" portions of tones, emphasizing the importance of pitch categorization in the perception of tonality.

Attraction. The theory also makes predictions about tonal attraction. In areas where resonance regions overlap (e.g., Figure 4B), more stable resonances overpower less stable ones, such that the instantaneous frequency of the population in the overlap region is attracted toward the frequency of the more stable resonance. To understand the implications for tonal attraction, a nonlinear

---

[1] Operationalization of "close" frequency as 1% is somewhat arbitrary, and different choices result in different resonance regions for the weaker resonances, changing the predictions slightly, but not altering the basic results.

oscillator network was simulated using internal connectivity among oscillators that reflected the structure of the major scale. The simulation was based on a two-layer network, modeling a cochlear transformation followed by a neural transformation. This minimal model provided a simple example in which tonal attraction can be observed. The first layer parameters were $\alpha = -.01$; $\omega = 2\pi$; $\beta_1 = \beta_2 = \beta_n = -1$; $\delta_1 = \delta_2 = \delta_n = 0$; $\varepsilon = 0.1$ and $\tau_i = 1/f_i$, where f_i is the frequency of each oscillator in Hz, similar to the first simulation, but without frequency detuning. The frequencies of the network were distributed along a logarithmic frequency gradient, with 120 oscillators per octave, spanning two octaves. A Gaussian kernel modeled local basilar membrane coupling (cf. Kern & Stoop, 2003).

The parameters of the second network were set to $\tau_i = 1/f_i$ where f_i is the frequency of each oscillator in Hz, and $\alpha = -0.4$; $\beta_1 = 1.2$; $\beta_2 = \beta_n = -1$; $\delta_1 = -0.01$; $\varepsilon = 0.75$. All other parameters were set to zero. Again, the frequencies of the network were distributed along a logarithmic frequency gradient, with 120 oscillators per octave, spanning two octaves. The center frequency of both networks was chosen to match middle C. Afferent connectivity from the cochlear network was one-to-one, with oscillators of the cochlear network stimulating oscillators of the neural network according to frequency. In the second network internal connectivity was constructed to reflect learning of the ET scale, as described above. Each oscillator was connected to the others that are nearby in frequency, as well as to those whose eigenfrequencies approximated the frequency ratios of the scale. The main feature of this simulation is that the parameters of the second (neural) network are chosen to be near a degenerate Hopf bifurcation, also known as a Bautin bifurcation (Guckenheimer & Kuznetsov, 2007). For this reason, the nonlinear coupling allows amplitude peaks to self-stabilize at the frequencies of stimulation, such that after the stimulus is removed, the peaks remain. This behavior is seen in Panels D, E, and F of Figure 5. Individual peaks interact with one another as well. Interactions in the gradient frequency network are complex, and a complete analysis is beyond the scope of this chapter. However, self-stabilizing amplitude peaks have been observed and analyzed for single-frequency oscillator networks with nonlinear coupling near a Bautin bifurcation (Drover & Ermentrout, 2003).

The network was stimulated with a C-major triad (the notes C, E, and G), which was followed after a delay by a leading tone (the note B; Figure 5A & B), and the instantaneous frequencies (Panel C) and amplitudes (Panels D, E, F) of the oscillators in the network were measured. The stimulus was prepared using Finale Notepad Plus 2005a, saved as a MIDI file, and rendered to a digital audio file as pure tones. After stimulation with the triad, a dynamic field self-stabilized to embody a set of resonant frequencies that was consistent with the prior stimulation, embodying a memory of the stimulus. Immediately before introduction of the leading tone ($t = 1.0$), stable amplitude peaks corresponding to the populations of oscillators surrounding C, E and G are observed (Panel D), and the instantaneous frequencies of the three oscillators at C, E and G also appear stable (Panel C, $t = 1.0$). After the leading tone is introduced, a corresponding

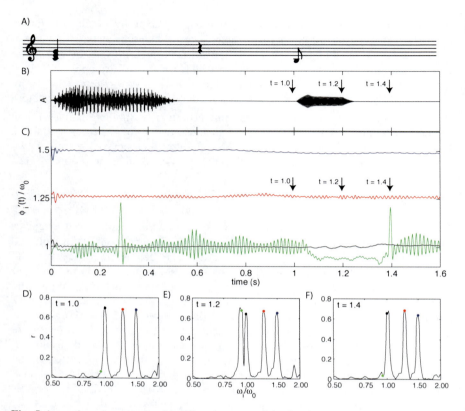

Fig. 5 Attraction. A network with ET scale connectivity is stimulated with a C major triad (C, E, G; scale degrees 1, 3, 5), followed after a delay with a leading tone (B, scale degree 7). A) Musical notation. B) Stimulus waveform. C) Instantaneous frequency of four oscillators (out of 241 in the network) closest in eigenfrequency to scale degrees 7 (green), 1 (black), 3 (red), 5 (blue). After the triad is silenced, the dynamic field remains stable (D) with peaks corresponding to scale degrees 1, 3, and 5. When stimulation at scale degree 7 begins, a corresponding peak arises (E) and its frequency stabilizes (green). After stimulation ceases, the oscillator with eigenfrequency near scale degree 7 loses stability, the peak dies away (F), and its instantaneous frequency is attracted toward the tonic.

amplitude peak is observed (Panel E) and the instantaneous frequency is stabilized by the external stimulus (Panel C; $t = 1.2$, green). The important observation is that when the external stimulus is removed, this oscillation loses stability ($t = 1.4$) and its frequency is attracted toward the note C, the tonic frequency. This network behavior predicts a physical correlate for the perceived attraction of the leading tone to the tonic, and in general for expectation of what should follow in a tonal context.

Stability. The next analysis asked whether theoretical stability of higher order resonances could predict perceived stability of tones (Krumhansl, 1990). For the

A Dynamical Systems Approach to Musical Tonality

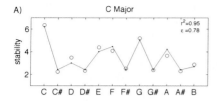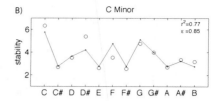

Fig. 6 Comparison of theoretical stability predictions and human judgments of perceived stability for two Western modes. A) C major, B) C minor.

stability analysis, frequency ratios were fixed according to the previous analysis of learning (Figure 4B). It was further assumed that all the non-zero c were equal, effectively eliminating one free parameter (although in principle, the coupling strength, c, could be different for each resonance as a result of learning). Relative stability of each resonance was predicted by $\varepsilon^{(k+m-1)/2}$, where k and m are the numerator and denominator of the frequency ratio, respectively, and $0 \leq \varepsilon \leq 1$ is a parameter that controls nonlinear gain (Hoppensteadt & Izhikevich, 1997). The analysis assumed that each tone listeners heard as part of the context sequence was stabilized in the network, and those that were not heard were not stabilized. This assumption reflects the behavior of the network simulated in the previous analysis. Thus, each context tone received a stability value of $\varepsilon^{(k+m-1)/2}$, and those that did not occur in the context sequence received a stability value of 0. For major and minor Western modes, the parameter ε was chosen to maximize the correlation (r^2, different from oscillator amplitude, r, used previously) with the stability ratings of listeners. This provides a single parameter fit to the perceptual data on stability, shown in Figure 6.

Predicted stability matched the perceptual judgments well (C major: $r^2 = .95$, $p < .0001$, $\varepsilon = 0.78$; C minor: $r^2 = .77, p < .001$, $\varepsilon = 0.85$), as shown in Figure 6. In other words, the theoretical stability of higher-order resonances of nonlinear oscillators predicts empirically measured tonal stability for major and minor tonal contexts. This result is significant because it does not depend on the statistics of tone sequences, but instead it predicts the statistics of tone sequences, which are known to be highly correlated with stability judgments (Krumhansl, 1990).

5 Discussion

While the properties of nonlinear resonance predict the main perceptual features of tonality well, this theory makes two additional significant predictions: 1) that nonlinear resonance should be found in the human auditory system and 2) that animals with auditory systems similar to humans may be sensitive to tonal relationships. Recently, evidence has been found in support of both predictions.

Helmholtz (1863) described the cochlea as a time-frequency analysis mechanism that decomposes sounds into orthogonal frequency bands for further analysis by the central auditory nervous system. Von Bekesey (1960) observed that human cadaver cochlear responses behave linearly over the range of physiologically relevant sound intensities. However, the weakest audible sounds impart energy per cycle no greater than that of thermal noise (Bailek, 1987), and the system operates over a range of intensities that spans at least 14 orders of magnitude. Moreover, laser-interferometric velocimetry performed on living, intact cochleae has revealed exquisitely sharp mechanical frequency tuning (Ruggero, 1992). Recent evidence, including the discovery of spontaneous otoacoustic emissions (Kemp, 1979), suggest that the sharp mechanical frequency tuning, exquisite sensitivity and operating range of the cochlea can be explained by critical nonlinear oscillations of hair cells (Choe, Magnasco, & Hudspeth, 1998). Thus, the cochlea performs an active, nonlinear transformation, using a network of locally coupled outer-hair cell oscillators, each tuned to a distinct eigenfrequency.

There is a growing body of evidence consistent with nonlinear oscillation in the central auditory system as well. In mammals, action potentials phase-lock to both fine time structure and temporal envelope modulations at many different levels in the central auditory system, including cochlear nucleus, superior olive, inferior colliculus (IC), thalamus and A1 (Langner, 1992), and recent evidence points to a key role for synaptic inhibition in maintaining central temporal representations. Hyperpolarizing inhibition is phase-locked to the auditory stimulus and has been shown to adjust the temporal sensitivity of coincidence detector neurons (Grothe, 2003), while stable pitch representation in the IC may be the result of a synchronized inhibition originating from the ventral nucleus of the lateral lemniscus (Langner, 2007). Moreover, neurons in the IC of the gerbil have been observed to respond at harmonic ratios (3:2, 2:1, 5:2, etc.) with the temporal envelope of the stimulating waveform (Langner, 2007). *Pollimyrus*, an fish that lacks a sophisticated peripheral structure for mechanical frequency analysis, has modulation-rate selective cells in the auditory midbrain that receive both excitatory and inhibitory input, and are well described as nonlinear oscillators (Large & Crawford, 2002). Finally, residue pitch shift – a central auditory phenomenon – is consistent with 3-frequency resonances of nonlinear oscillators, making nonlinear resonance viable as a candidate for the neural mechanism of pitch perception in humans (Cartwright, Gonzalez, & Piro, 1999).

If key aspects of tonality depend directly on auditory physiology, one would predict that non-human animals might be sensitive to certain tonal relationships. Wright et al. tested two rhesus monkeys for octave generalization in eight experiments by transposing 6- and 7-note musical passages by an octave and requiring *same* or *different* judgments (Wright, Rivera, Hulse, Shyan, & Neiworth, 2000). The monkeys showed complete octave generalization to childhood songs (e.g., "Happy Birthday") and tonal melodies (from a tonality algorithm). They showed no octave generalization to random-synthetic melodies, atonal melodies, or individual notes. Takeuchi's Maximum Key Profile Correlation measure of tonality, based on human tonality judgments (Takeuchi, 1994) accounted for 94 percent of the variance in the monkey data. These results provide evidence that tonal melodies retain their identity when transposed with whole octaves, as they do for humans. Adult listeners can recognize transpositions of tonal but not atonal melodies (Dowling & Fujitani, 1971). Preverbal infants can recognize transposed tonal melodies (Trehub, Morrongiello, & Thorpe, 1985), and melody identification is nearly perfect for octave (2:1 ratio) transpositions, even for novel melodies (Massaro, Kallman, & Kelly, 1980), as is also the case for macaques.

Zuckerkandl (1956) argued that the dynamic quality of musical tones "...makes melodies out of successions of tones and music of acoustical phenomena." The current approach predicts that the perceived dynamics of tonal organization arise from the physics of nonlinear resonance. Thus, nonlinear resonance may provide the neural substrate for a substantive musical universal. In some ways, this idea is similar to the concept of universal grammar in linguistics (Prince & Smolensky, 1997). However, in the case of music, these perceptual universals are predicted by universal properties of nonlinear resonance, offering a direct link to neurophysiology. Learning would alter connectivity to establish different resonances, and different tonal relationships. Higher-order resonances may give rise to dynamic tonal fields in the central nervous system, with localized patterns of activation self-stabilizing to embody the musical system of the listener's culture. Musical melodies would be perceived in relation to the tonal field, creating a dynamic context within which perception of tone sequences takes place.

Acknowledgments. This research was supported by NSF CAREER Award BCS-0094229 and NSF grant BCS-1027761. The author would like to thank Justin London, Carol Krumhansl, John Iversen, Frank Hoppensteadt, and Eugene Izhikevich for insightful comments on an earlier draft of this chapter. Thanks also to Felix Almonte, Marc Velasco and Nicole Flaig for many interesting discussions, and for their invaluable help with this manuscript.

References

Bailek, W.: Physical limits to sensation and perception. Annual Review of Biophysics and Biophysical Chemistry 16, 455–478 (1987)

Bharucha, J.J.: Anchoring effects in music: The resolution of dissonance. Cognitive Psychology 16, 485–518 (1984)

Bharucha, J.J., Stoeckig, K.: Reaction-Time and Musical Expectancy - Priming of Chords. Journal of Experimental Psychology-Human Perception and Performance 12, 403–410 (1986)

Burns, E.M.: Intervals, scales, and tuning. In: Deustch, D. (ed.) The Psychology of Music, pp. 215–264. Academic Press, San Diego (1999)

Burns, E.M., Campbell, S.L.: Frequency and frequency ratio resolution by possessors of relative and absolute pitch: Examples of categorical perception? Journal of the Acoustical Society of America 96, 2704–2719 (1994)

Camalet, S., Duke, T., Julicher, F., Prost, J.: Auditory sensitivity provided by self tuned critical oscillations of hair cells. Proceedings of the National Academy of Sciences 97, 3183–3188 (1999)

Cartwright, J.H.E., Gonzalez, D.L., Piro, O.: Nonlinear Dynamics of the Perceived Pitch of Complex Sounds. Physical Review Letters 82, 5389–5392 (1999)

Choe, Y., Magnasco, M.O., Hudspeth, A.J.: A model for amplification of hair-bundle motion by cyclical binding of Ca2+ to mechanoelectrical-transduction channels. Proceedings of the National Academy of Sciences 95, 15321–15336 (1998)

Cuddy, L.L., Lunney, C.A.: Expectancies generated by melodic intervals: Perceptual judgements of melodic continuity. Perception & Psychophysics 57, 451–462 (1995)

Dowling, W.J.: Scale and contour: Two components of a theory of memory for melodies. Psychological Review 85, 341–354 (1978)

Dowling, W.J., Fujitani, D.S.: Contour, interval and pitch recognition in memory for melodies. Journal of the Acoustical Society of America 49, 524–531 (1971)

Drover, J.D., Ermentrout, B.: Nonlinear coupling near a degenerate Hopf (Bautin) Bifurcation. SIAM Journal On Applied Mathematics 63, 1627–1647 (2003)

Eerola, T., Toiviainen, P.: Finnish Folksong Database (2004), https://www.jyu.fi/hum/laitokset/musiikki/en/research/coe/materials/collection_download (Retrieved October 13, 2010)

Grothe, B.: New roles for synaptic inhibition in sound localization. Nature Reviews Neuroscience 4, 540–550 (2003)

Grothe, B., Klump, G.M.: Temporal processing in sensory systems. Current Opinion in Neurobiology 10, 467–473 (2000)

Guckenheimer, J., Kuznetsov, Y.A.: Bautin bifurcation. Scholarpedia 2, 1853 (2007)

Helmholtz, H.L.F.: On the sensations of tone as a physiological basis for the theory of music. Dover Publications, New York (1863)

Hoppensteadt, F.C., Izhikevich, E.M.: Synaptic organizations and dynamical properties of weakly connected neural oscillators I: Analysis of a canonical model. Biological Cybernetics 75, 117–127 (1996a)

Hoppensteadt, F.C., Izhikevich, E.M.: Synaptic organizations and dynamical properties of weakly connected neural oscillators II: Learning phase information. Biological Cybernetics 75, 126–135 (1996b)

Hoppensteadt, F.C., Izhikevich, E.M.: Weakly Connected Neural Networks. Springer, Heidelberg (1997)

Kameoka, A., Kuriyagawa, M.: Consonance theory part II: Consonance of complex tones and its calculation method. Journal of the Acoustical Society of America 45, 1460–1471 (1969)

Kemp, D.T.: Evidence of mechanical nonlinearity and frequency selective wave amplification in the cochlea. European Archives of Oto-Rhino-Laryngology 224, 370 (1979)

Kern, A., Stoop, R.: Essential role of couplings between hearing nonlinearities. Physical Review Letters 91, 128101–128104 (2003)

Knopoff, L., Hutchinson, W.: Entropy as a measure of style: The influence off sample length. Journal of Music Theory 27, 75–97 (1983)

Krumhansl, C.L.: Cognitive foundations of musical pitch. Oxford University Press, NY (1990)

Krumhansl, C.L., Kessler, E.J.: Tracing the dynamic changes in perceived tonal organization in a spatial representation of musical keys. Psychological Review 89(4), 334–368 (1982)

Langner, G.: Periodicity coding in the auditory system. Hearing Research 60, 115–142 (1992)

Langner, G.: Temporal processing of periodic signals in the auditory system: Neuronal representation of pitch, timbre, and harmonicity. Z Audiol. 46, 21–80 (2007)

Large, E.W., Almonte, F., Velasco, M.: A canonical model for gradient frequency neural networks. Physica D: Nonlinear Phenomena 239(12), 905–911 (2010)

Large, E.W., Crawford, J.D.: Auditory temporal computation: Interval selectivity based on post-inhibitory rebound. Journal of Computational Neuroscience 13, 125–142 (2002)

Larson, S.: Musical Forces and Melodic Expectations: Comparing Computer Models and Experimental Results. Music Perception 21, 457–498 (2004a)

Lerdahl, F.: Tonal Pitch Space. Oxford University Press, New York (2001)

Loosen, F.: Intonation of solo violin performance with reference to equally temepred, Pythagorean and just intonations. Journal of the Acoustical Society of America 93, 25–539 (1993)

Massaro, D.W., Kallman, H.J., Kelly, J.L.: The role of tone height, melodic contour, and tone chroma in melody recognition. Journal of Experimental Psychology: Human Learning and Memory 6, 77–90 (1980)

Plomp, R., Levelt, W.J.M.: Tonal consonance and critical bandwidth. Jounal of the Acoustical Society of America 38, 548–560 (1965)

Prince, A., Smolensky, P.: Optimality: From Neural Networks to Universal Grammar. Science 275, 1604–1610 (1997)

Ruggero, M.A.: Responses to sound of the basilar membrane of the mamalian cochlea. Current Opinion in Neurobiology 2, 449–456 (1992)

Smith, J.D., Nelson, D.G., Grohskopf, L.A., Appleton, T.: What child is this? What interval was that? Familiar tunes and music perception in novice listeners. Cognition 52, 23–54 (1994)

Tillmann, B., Bharucha, J.J., Bigand, E.: Implicit learning of tonality: A self-organizing approach. Psychological Review 107, 885–913 (2000)

Trehub, S.E., Morrongiello, B.A., Thorpe, L.A.: Children's perception of familiar melodies: The role of interval contour and key. Psychomusicolgy 5, 39–48 (1985)

von Bekesy, G.: Experiments in Hearing. McGraw-Hill Book Co., New York (1960)

Wiggins, S.: Introduction to Applied Nonlinear Dynamical Systems and Chaos. Springer, New York (1990)

Wilson, H.R., Cowan, J.D.: A mathematical theory of the functional dynamics of cortical and thalamic nervous tissue. Kybernetik 13, 55–80 (1973)

Wright, A.A., Rivera, J.J., Hulse, S.H., Shyan, M., Neiworth, J.J.: Music perception and octave generalization in Rhesus monkeys. Journal of Experimental Psychology: General 129, 291–307 (2000)

Zuckerkandl, V.: Sound and Symbol: Music and the External World (WR Trask, Trans.). Princeton University Press, Princeton (1956)

Author Index

Calvin, Sarah 91

Daffertshofer, Andreas 35
Danion, Frédéric 115
Ditzinger, Thomas 179

Fuchs, Armin 1

Hock, Howard S. 151
Huys, Raoul 69

Jirsa, Viktor K. 91

Lancia, Leonardo 135
Large, Edward W. 193

Nguyen, Noël 135

Schöner, Gregor 151

Tuller, Betty 135

Vallabha, Gautam K. 135